SpringerBriefs in Mathematics

SpringerBriefs in Mathematics showcases expositions in all areas of mathematics and applied mathematics. Manuscripts presenting new results or a single new result in a classical field, new field, or an emerging topic, applications, or bridges between new results and already published works, are encouraged. The series is intended for mathematicians and applied mathematicians. All works are peer-reviewed to meet the highest standards of scientific literature.

More information about this series at http://www.springer.com/series/10030

Alfonso Zamora Saiz • Ronald A. Zúñiga-Rojas

Geometric Invariant Theory, Holomorphic Vector Bundles and the Harder-Narasimhan Filtration

 Springer

Alfonso Zamora Saiz [ID]
Applied Mathematics
Technical University of Madrid
Madrid, Spain

Ronald A. Zúñiga-Rojas [ID]
School of Mathematics
University of Costa Rica
San José, Costa Rica

ISSN 2191-8198   ISSN 2191-8201 (electronic)
SpringerBriefs in Mathematics
ISBN 978-3-030-67828-9   ISBN 978-3-030-67829-6 (eBook)
https://doi.org/10.1007/978-3-030-67829-6

Mathematics Subject Classification: 14-XX, 14-02, 14A30, 14D20, 14D21, 14H60

This Springer imprint is published by the registered company Springer Nature Switzerland AG
The registered company address is: Gewerbestrasse 11, 6330 Cham, Switzerland

Preface

On March 2016, the authors gave a series of talks at Universidad de Costa Rica on Geometric Invariant Theory (GIT), mostly devoted to understand the main features of the subject and being able to face its main application, the construction of moduli spaces. With the purpose of this short course, a series of notes were written, which, some years later, constitute this monograph.

This text intends to present the main elements of Geometric Invariant Theory, a technique to obtain quotients in algebraic geometry with good properties, through various examples. We go from the classical Hilbert classification of binary forms to the construction of the moduli space of semistable holomorphic vector bundles by Mumford and others, or Hitchin's theory on Higgs bundles. Although many excellent books already exist in the literature, the point of view taken here pays special attention to unstable objects in moduli problems, the ones left out from the construction of moduli spaces. The notion of the Harder-Narasimhan filtration as the main tool to handle them and its relationship with GIT quotients provide interesting new calculations in several problems. As applications, surveys of research results of the authors on correspondences between Harder-Narasimhan filtrations with the GIT picture and stratifications of the moduli space of Higgs bundles are provided.

The book tries to be self-contained and avoid generalizations which can be encountered in the references. Most of the proofs are omitted, and many are just sketched, in an attempt to show the ideas more than being completely precise in a mathematical sense. The language of algebraic varieties and vector bundles is kept whenever possible, despite the fact that the theory is developed in the framework of schemes and sheaves. Also, much of the theory and results presented in here holds in higher dimensional varieties, although we restrict the exposition to compact Riemann surfaces or, algebraically speaking, smooth complex projective curves.

The book is devoted to graduate students and researchers who want to approach Geometric Invariant Theory in moduli constructions, how the notion of stability appears in different contexts, and the important role played by the Harder-Narasimhan filtration in the study of unstable objects. Prerequisites to understand the material in this book are general courses on algebraic geometry and differential geometry. The reader willing to properly understand the subject should combine this material with the books and articles quoted in the references.

We wish to thank a number of people for their knowledge, expertise, and stimulating conversations towards completing this work: Luis Álvarez-Cónsul, Héctor Barge, Óscar García-Prada, Tomás L. Gómez, Peter Gothen, Carlos Florentino, Jorge Herrera de la Cruz, Alessia Mandini, Javier Martínez, Peter Newstead, André Oliveira, Milena Pabiniak, Ignacio Sols. We give special thanks to the manuscript referees and editors for carefully reviewing the text and suggesting insightful ideas.

We also thank the support of Universidad de Costa Rica and its Escuela de Matemática, specifically Centro de Investigaciones Matemáticas y Metamatemáticas (CIMM) and Oficina de Asuntos Internacionales y de Cooperación Externa (OAICE) through Programa Académicos Visitantes (PAV), for their support and the invitation to give a course in March 2016, where part of this material was covered, and to the organizers of XXII SIMMAC held in Costa Rica in February 2020, where part of this work was done, for their hospitality.

The first author is supported by projects MTM2016-79400-P and PID2019-108936GB-C21 of the Spanish government. The second author is supported by Escuela de Matemática and CIMM from Universidad de Costa Rica through projects 820-B5-202 and 820-B8-224.

Madrid, Spain   Alfonso Zamora Saiz

San José, Costa Rica   Ronald A. Zúñiga-Rojas
October 2020

Contents

List of Symbols

Here we include a list of symbols used in this book.

$M^{n \times n}(k)$	square $n \times n$ matrices over k
A^t	transpose matrix
A^*	adjoint matrix
$\det(A)$	determinant of A
$\text{tr}(A)$	trace of A
I_n	identity $n \times n$ matrix
G	Algebraic or Lie group
$\mathfrak{g}$	Lie algebra
K	compact Lie group
$K_{\mathbb{C}}$	complexified Lie group of K
$\text{GL}(n, k)$	general linear group
$\text{SL}(n, k)$	special linear group
$\text{U}(n)$	unitary group
$\text{SU}(n)$	special unitary group
$\text{O}(n, k)$	orthogonal group
$\text{SO}(n, k)$	special orthogonal group
$\text{Sp}(n, k)$	symplectic group
G_d	product of general linear groups with dimension vector d
Δ	subgroup of diagonal matrices of G_d
G_x	stabilizer of a point x in a group action
$O(x)$	orbit of x under the action of a group
$[v, w]$	Lie bracket
$\mathfrak{u}(n)$	Lie algebra of the unitary group
$\mathfrak{so}(n, \mathbb{R})$	Lie algebra of the special real orthogonal group
$\text{ad}(X)(\cdot)$	adjoint representation
$\text{Irr}\, G$	set of irreducible representations of G
χ	character of a group G
S^n	n-sphere

$\mathbb{A}^n_k$	affine n-dimensional space over k
$\mathbb{P}^n_k$	projective n-dimensional space over k
$\mathrm{Gr}(r, n)$	Grassmannian of r-planes on k^n
$Z(S)$	algebraic set defined by S
$I(X)$	ideal of a variety X
$I_h(X)$	homogeneous ideal of a variety X
$A(X)$	coordinate ring of an affine variety X
$A_h(X)$	homogeneous coordinate ring of a projective variety X
$\hat{X}$	affine cone of X
$A(X)^G$	ring of invariants of X by the action of G
$\mathrm{Spec}(R)$	spectrum of a ring R
$\mathrm{Proj}(R)$	projective spectrum of a graded ring R
$\mathfrak{p}$	prime ideal of a ring
$\mathfrak{q}$	homogeneous prime ideal of a graded ring
$\mathcal{U} = \{U_i\}_{i \in I}$	open covering of a space
$\mathcal{O}_X$	structure sheaf of X or sheaf of holomorphic functions on X
$\mathcal{E}_X$	sheaf of smooth functions on X
$\mathcal{E}^k_X$	sheaf of smooth differential k-forms on X
$\mathcal{E}^{p,q}_X$	sheaf of smooth differential complex-valued (p, q)-forms on X
Ω^k_X	sheaf of holomorphic differential k-forms on X
$\underline{G}$	locally constant sheaf with group G
$\mathbb{E}$	smooth complex vector bundle
E	holomorphic vector bundle
$\mathrm{rk}(E)$	rank of a vector bundle E
$\deg(E)$	degree of a vector bundle E
$\mathrm{Ker}\,\alpha$	kernel of α
$\mathrm{Im}\,\alpha$	image of α
g_{ij}	transition functions
$H^i(X, E)$	ith-group of cohomology of X with coefficients in the sheaf E
$h^i(X, E)$	dimension of the ith-cohomology group
$E \oplus F$	direct sum of vector bundles
$E \otimes F$	tensor product of vector bundles
$E^\vee$	dual vector bundle
L^{-1}	inverse line bundle
$\wedge^k E$	kth-exterior power of E
$\mathrm{Hom}(E, F)$	homomorphisms from E to F
$\mathrm{Hom}^G(E, F)$	G-invariant homomorphisms from E to F
$\det(E)$	determinant of E
TX	tangent bundle to X
$TX^\vee$	cotangent bundle to X
ω_X	dualizing sheaf, canonical line bundle

$\mathrm{Pic}(X)$, $\mathrm{Pic}^d(X)$	Picard group and degree d component
$\mathcal{O}_X(1)$	twisting sheaf of Serre
$\chi(E)$	Euler characteristic of E
$P_E(m)$	Hilbert polynomial of E
D	divisor
$|D|$	complete linear system
g	genus of a compact Riemann surface X
$c_i(E)$	ith-Chern class of E
$c(E)$	total Chern class of E
$f_*(E)$	push-forward of E by f
$f^{-1}(E)$	inverse image of E by f
$f^*(E)$	pull-back of E by f
$\mathcal{E}_X^k(E)$	sheaf of smooth differential k-forms on X with values in E
$\mathcal{E}_X^{p,q}(E)$	sheaf of smooth differential complex-valued (p,q)-forms on X with values in E
$\Omega_X^k(E)$	sheaf of holomorphic differential k-forms on X with values in E
$X /\!/ G$	GIT quotient of X by G
X^s, X^{ss}, X^{ps}	stable, semistable, polystable points of X
ρ	1-parameter subgroup of G
$\mathcal{R}(G)$	set of 1-parameter subgroups of G
$\vartheta(x, \rho)$	weight of the action of the 1-parameter subgroup ρ over x
$\mu(E)$	slope of E
Quot	Quot-scheme
Quot^G	G-invariant Quot-scheme
$\mathcal{M}(r, L)$, $\mathcal{M}_s(r, L)$	moduli spaces of semistable and stable holomorphic vector bundles of rank r and determinant L
$\mathcal{M}(r, d)$, $\mathcal{M}_s(r, d)$	moduli spaces of semistable and stable holomorphic vector bundles of rank r and degree d
$\mathrm{gr}(E)$	graded object of E
$\vec{\mu}(E)$	Harder-Narasimhan type of E
ω	symplectic form
$\mathrm{Symp}(X, \omega)$	symplectomorphisms of X
$\mathrm{Diff}(X)$	diffeomorphisms of X
$\mathrm{Vect}(X)$, $\mathrm{Vect}^s(X)$, $\mathrm{Vect}^H(X)$	vector fields, symplectic and Hamiltonian vector fields of X
$\mathcal{L}_v \zeta$	Lie derivative of ζ along v
$\iota_v \zeta$	interior product of ζ with v
$d\zeta$	exterior derivative of ζ
ψ	moment map
$\|\psi\|$	moment map square
$\Psi_{\hat{x}}$	Kempf-Ness function

$\Xi_x(g)$ moment map square on the orbit of x

$\Phi_x(\rho)$ Kempf function

$\bar{\partial}$ Cauchy's differential operator

$\bar{\partial}_E$ Dolbeault's differential operator

$\mathcal{A}^{0,1}(\mathbb{E}), \mathcal{A}_s^{0,1}(\mathbb{E}), \mathcal{A}_{ps}^{0,1}(\mathbb{E})$ spaces of holomorphic, stable, and polystable structures $\bar{\partial}_E$ on $\mathbb{E}$

$\mathcal{G}, \mathcal{G}^{\mathbb{C}}$ gauge group and complex gauge group

$\mathcal{N}(r, d), \mathcal{N}_s(r, d)$ Dolbeault's moduli spaces of semistable and stable holomorphic vector bundles of rank r and degree d

$\mathrm{GCD}(r, d)$ greatest common divisor of r and d

∇ connection

Θ_∇ curvature of a connection

$\pi_1(X, x_0)$ fundamental group of X with base point x_0

$\pi_n(X)$ nth-homotopy group of X

$\widetilde{X}$ covering of X

$\mathcal{R}(r, d), \mathcal{R}_{irr}(r, d)$ moduli spaces of degree d unitary and irreducible unitary representations

(E, φ) Higgs bundle

$\mathcal{H}, \mathcal{H}_s, \mathcal{H}_{ps}$ spaces of configurations, stable and polystable configurations of Higgs bundles

$\mathcal{M}^H(r, d), \mathcal{M}_s^H(r, d)$ moduli spaces of polystable and stable Higgs bundles of rank r and degree d

$\mathcal{C}_h$ space of connections compatible with an hermitian metric h

$\mathcal{X}_0, \mathcal{X}^*$ spaces of solutions and irreducible solutions to the Hitchin equations

$\mathcal{M}^{YM}(r, d), \mathcal{M}_{irr}^{YM}(r, d)$ moduli spaces of solutions and irreducible solutions to the (Yang-Mills) Hitchin equations

$\mathcal{M}(r, d, L)$ moduli space of semistable Hitchin pairs

$\mathcal{R}^H(r, d), \mathcal{R}_{irr}^H(r, d)$ moduli spaces of degree d reductive and irreducible representations into $\mathrm{GL}(r, \mathbb{C})$

(E, ϕ) holomorphic pair

(E, τ) tensor

$\mathbb{P}(E)$ projectivization of the vector bundle E

Q quiver

Q_0, Q_1 vertices and arrows of a quiver Q

$\mathbb{Z}Q_0$ free abelian group generated by Q_0

$\mathrm{mod}\, kQ$ category of finite dimensional representations of Q over k

$\underline{\dim}M$ dimension vector of a representation M

$\mu_{(\theta,\sigma)}(M)$ (θ, σ)-slope of a representation M

$\mathcal{R}_d(Q)$ space of representations of a quiver Q with dimension vector d

$\mathcal{M}_d^{(\theta,\sigma)}(Q)$	moduli space of (θ,σ)-semistable representations of a quiver Q with dimension vector d
$\theta(\mathcal{F})$	stability function of a constellation $\mathcal{F}$
$r(\mathcal{F})$	rank, dimension of the negative part of a constellation $\mathcal{F}$
$\mu_\theta(\mathcal{F})$	θ-slope of a constellation $\mathcal{F}$
$U'_{\overrightarrow{\mu}(E)}$	Shatz stratum of Harder-Narasimhan type $\overrightarrow{\mu}(E)$
U_λ^+	Białynicki-Birula stratum of critical value c_λ
F_λ	critical submanifolds of the $\mathbb{C}^*$-action on the Higgs bundles moduli space
$\mathcal{M}^{H,k}(r,d)$	moduli space of semistable k-Higgs bundles
φ^k	k-Higgs field
$\mathcal{M}^\infty(r,d)$	direct limit of the moduli spaces of k-Higgs bundles

Chapter 1
Introduction

Moduli spaces are structures classifying objects under some equivalence relation, and many of these problems can be posed as quotients of an algebraic variety under the action of an algebraic group. The purpose of Geometric Invariant Theory (abbreviated GIT, [66, 67]) is to provide a way to define a quotient of the variety by the action of the group with an algebro-geometric structure, in the case when the group satisfies a condition: being reductive. This way, GIT results assure a likable structure for the quotient, giving a positive solution to the classification problem.

The solution to the quotient of a group action on a variety dates back to Hilbert's 14th problem. Understanding a variety by means of its ring of functions Nagata [69] proved that, for the action of a reductive group in an affine variety, the structure of the quotient is fully recovered by the ring of functions which are invariant by the action of the group: the ring of invariants of an affine variety is finitely generated; hence, it is the coordinate ring of an affine variety. Therefore we can define the quotient of the variety by the group as the affine variety associated to the ring of invariants. First thing to note is that the orbit space (i.e., the quotient space where each point corresponds to an orbit) is not separated, the reason why the ring of invariants identifies orbits to yield a Hausdorff quotient.

When taking the quotient of a projective variety by a group, there are other issues which have to be taken into account. In the projective case, the correspondence between affine varieties and coordinate rings has to be replaced by a correspondence between polarized varieties (i.e., varieties together with an ample line bundle providing an embedding into a projective space) and graded coordinate rings. This amounts to working on the affine cone of the projective variety, which is given by placing the line corresponding to each point of the variety inside the projective space (equivalent to blowing down the zero section in the total space of the ample line bundle), together with a linearization of the action, meaning a lift of the action to an action on the affine cone of the variety. This process of looking at the action of the group on a variety through its embedding into a projective space induces a game between the action orbits, where some of them play an unfortunate role and do not

A. Zamora Saiz, R. A. Zúñiga-Rojas, *Geometric Invariant Theory, Holomorphic Vector Bundles and the Harder-Narasimhan Filtration*, SpringerBriefs in Mathematics, https://doi.org/10.1007/978-3-030-67829-6_1

appear in the quotient by the group when embedded into the projective space, hence they have to be removed: this is how the notion of GIT stability appears.

Geometric Invariant Theory was developed by Mumford [66] as a theory to construct quotients of varieties, broadening the scope of Hilbert's problem. The name comes from the intention to provide a geometric meaning to former Hilbert's theory of invariants. After Narasimhan and Seshadri's celebrated theorem [70, 80] relating vector bundles and representations, GIT's major application was the construction of a projective variety classifying all holomorphic structures in a smooth bundle, which is called the moduli space of holomorphic vector bundles. This projective variety compactifies the analytical construction of a moduli space of differential operators defining holomorphic structures by Dolbeault [20], as well as a topological construction of a moduli space of representations. Based on Kirwan's work [59], there is another theory of symplectic quotients mirroring with the GIT picture, the Kempf-Ness theorem [57] being the link between stability from the GIT and symplectic points of view.

Extending representations of the unitary group to the general linear group we find the notion of Higgs bundle, first studied by Hitchin [45], as solutions of self-duality equations on the plane considering invariance through two trajectories, what gets conformal invariance and then solutions on compact Riemann surfaces with higher genus. These objects have turned to be a central element intertwining geometry, topology, and physics after the works of Atiyah-Bott [5], the generalization to higher dimension by Simpson [86], the study of its moduli space by Hausel [40], the moduli construction using GIT by Nitsure [72] or its Hitchin-Kobayashi correspondence in [21].

When constructing the moduli space of vector bundles, or other moduli spaces as for Higgs bundles, we impose a stability condition and declare certain objects to be stable, semistable, or polystable, such that when encoding the problem in the GIT framework, these notions match. Therefore, constructing a GIT quotient provides a moduli space classifying equivalence classes of orbits of semistable bundles, each one containing a polystable representative. This stability concept and the moduli theory leave outside of the quotient picture the non-semistable objects, called unstable. Unstable bundles are more complex: they carry bigger and different automorphism groups, and, therefore, the action of the group on them yields different stabilizers, which cause anomalies in the quotient (such as the jump phenomenon, see [71]), the reason why we have to remove them.

Hilbert-Mumford criterion is the tool to check GIT stability through the idea of 1-parameter subgroups, which are 1-dimensional subvarieties of the group accounting for the various features of separating orbits. GIT unstable points are localized as those contradicting the criterion, analogous to unstable bundles being the objects not verifying the stability condition for bundles. The result of Kempf [56] and the well-known Harder-Narasimhan filtration [37] show the two sides of this maximal principle. In the articles [28, 96], these two notions are put in correspondence showing that the notion of the Harder-Narasimhan filtration can be recovered from the GIT problem through the 1-parameter subgroup which maximally contradicts the Hilbert-Mumford criterion. This can, for example, help define Harder-Narasimhan sort of filtrations in examples where there is no previous knowledge of this [97]. Also, in moduli problems where the correspondence holds to some extent, or does not hold at all, studying it reveals important information about how the GIT problem captures stability and, particularly, the concept of unstability [9, 87]. Besides, this relationship is linked to symplectic and differential geometry in [23].

One of the main applications of the Harder-Narasimhan filtration is that it provides a way to stratify the moduli problem in terms of the numerical invariants of the object, in order to study its properties. For vector bundles, Shatz [83] constructs a locally closed stratification indexed by the data of the Harder-Narasimhan filtration, yielding many applications. For example in [49], together with the aforementioned idea of maximal unstability, it is given a moduli space structure for each strata. In the Higgs bundles case, these stratifications can be used to study the geometry and topology of the moduli space, see [8, 31, 40, 99, 100].

The monograph is intended to be self-contained and is organized as follows. In Chap. 2 we give the basic tools in geometry that we will need in the following. In Sect. 2.1 we review the basics about algebraic varieties, affine and projective, and actions by algebraic groups. In Sect. 2.2 we briefly expose the main technical tools to deal with the rest later, sheaf theory and sheaf cohomology, and define schemes which are the actual spaces where GIT is sustained. In Sect. 2.3 we study holomorphic vector bundles, line bundles which are rank one vector bundles, and its related notion of divisor, plus the idea of Chern classes and the Riemann-Roch theorem.

After that, in Chap. 3 we introduce Geometric Invariant Theory to construct quotients and the notion of GIT stability through the old Hilbert idea reformulated by Mumford, which is nowadays referred to as the Hilbert-Mumford criterion (Sects. 3.1 and 3.2). Section 3.3 links this theory to symplectic geometry, and Sect. 3.4 shows different examples through which these notions are visualized, the construction of the projective space and the Grassmannian variety as GIT or symplectic quotients, and the study of the classical problem (dating back to Hilbert) of classifying binary forms defining configurations of points on the projective line. Final Sect. 3.5 of this chapter is devoted to study and stratify those orbits in the quotient examples which are unstable, under some maximality idea coming from the Hilbert-Mumford criterion. This will be the key to understand the correspondences between the GIT setting and the Harder-Narasimhan filtration in Chap. 5, and the stratifications in unstability types in Chap. 6.

Chapter 4 goes over a sketch of the GIT construction of the moduli space of holomorphic vector bundles in Sect. 4.1 and presents the Harder-Narasimhan filtration in Sect. 4.2. In Sect. 4.3 we recover other related constructions of the moduli space of vector bundles: Dolbeault moduli space of differential operators and Narasimhan-Seshadri moduli of representations of the fundamental group of a compact Riemann surface. Then, in Sect. 4.4 we state the idea of Higgs bundles started by Hitchin, and their moduli space from the algebraic and analytical points of view.

Chapter 5 includes results on correspondences between Harder-Narasimhan filtrations and maximally destabilizing 1-parameter subgroups in the Hilbert-Mumford criterion for unstable objects in different moduli problems. Section 5.1 contains the main correspondence for the moduli of holomorphic vector bundles over a compact Riemann surface, plus some generalizations to other bundle-related problems: holomorphic pairs, Higgs bundles, and tensors. Section 5.2 explores this correspondence in a different environment which are representations of a quiver, or a directed graph, into the category of finite dimensional vector spaces. Finally, Sect. 5.3 extends the quivers problem to an infinitely many parameters universe: constellations.

To conclude, Chap. 6 shows results about stratifications on the moduli space of Higgs bundles. In terms of the Harder-Narasimhan filtration of the underlying vector bundle, Shatz provides a stratification of the moduli space, which is recalled in Sect. 6.1. For rank two, this stratification is proved in [40] to coincide with the Białynicki-Birula stratification given by an action in the moduli space, as explained in Sect. 6.2. Section 6.3 extends these ideas to rank three Higgs bundles, and Sect. 6.4 makes use of the stratifications to study the topology of the moduli space through its homotopy groups.

There are excellent surveys, articles, and books about many topics covered in this monograph. For a treatment on Geometric Invariant Theory, the reader can directly read the seminal work of Mumford [66] or the extension with Fogarty and Kirwan [67]. The book by Newstead [71], published later, is also considered a cornerstone for understanding GIT and its application to the main classification problems addressed in here. The more recent books by Mukai [65] and Schmitt [79] are also main references in the subject. Kirwan's thesis [59] adds the theory of symplectic quotients to this picture. Also, Woodward and Thomas' surveys [89, 93] face this topic from symplectic-differential and algebraic flavors, respectively.

The starting point in the moduli history is the original paper by Riemann [75]. Crucial works in the history of moduli spaces of vector bundles are classifications of bundles on the projective line and an elliptic curve by Grothendieck [33] and Atiyah [4], respectively. To generalize the problem to higher dimensional varieties, there is Gieseker [24] for surfaces and Maruyama [61] on higher dimension. For Higgs bundles, we need to refer to Hitchin's foundational paper [45], Nitsure's algebraic construction [72], and Simpson's generalization to higher dimension in the so-called non-abelian Hodge Theory [85, 86].

For the notion of GIT stability and references related to the Hilbert-Mumford criterion, we need to quote old Hilbert's paper [44] stating this criterion for the very first time in the binary forms classification, and the theorem by Kempf [56] on maximal 1-parameter subgroups. An exciting theory of unstability in the general framework of stacks is presented by Halpern-Leistner [35, 36].

To apprehend the algebraic geometry necessary background to understand the constructions discussed in this book, we refer the reader to Hartshorne [38] or Shafarevich [82]. For an algebraic geometry treatment with a gauge theoretical flavor, we have Wells [92]. Modern sheaf theoretical treatments of moduli spaces can be learned from Huybrechts-Lehn [53]. The tools from differential geometry can be covered by Warner [91] and differential and algebraic topology by Bott-Tu [11]. Theory about Lie groups, algebras, and representations can be found in the books [39, 50], as well as group actions in [1, 14].

Chapter 2
Preliminaries

In this chapter, we collect all the necessary background to follow the further discussion on geometric invariant theory and moduli spaces. First we recover the notions of algebraic (affine and projective) variety and actions of algebraic groups, which will be the features in GIT quotients. Then we include a brief summary of sheaves, cohomology, and schemes, because those are the ingredients to develop this theory in full generality. Finally, essentials about holomorphic vector bundles, line bundles, and divisors are discussed.

2.1 Algebraic Varieties and Groups

Let us present notations, definitions, and properties of algebraic varieties and actions of algebraic groups on them.

2.1.1 Algebraic Varieties

Let k be an algebraically closed field and denote by $\mathbb{A}^n_k$ the n-dimensional affine space over k. Let Σ be a set of polynomials in the ring $k[X_1, \ldots, X_n]$ and define the subset of the affine space where all elements of Σ take the value zero as

$$Z(\Sigma) = \left\{ x \in \mathbb{A}^n_k : f(x) = 0 \text{ for all } f \in \Sigma \right\}.$$

Definition 2.1 A subset $Y \subset \mathbb{A}^n_k$ is an *algebraic set* if there exists a set of polynomials $\Sigma \subset k[X_1, \ldots, X_n]$ such that $Y = Z(\Sigma)$.

© The Author(s), under exclusive license to Springer Nature Switzerland AG 2021
A. Zamora Saiz, R. A. Zúñiga-Rojas, *Geometric Invariant Theory, Holomorphic Vector Bundles and the Harder-Narasimhan Filtration*, SpringerBriefs in Mathematics, https://doi.org/10.1007/978-3-030-67829-6_2

By declaring all algebraic sets $Z(\Sigma)$ to be closed, we endow an affine space with the *Zariski topology*. Recall that a subset Y of a topological space is irreducible if it cannot be described as the union $Y = Y_1 \cup Y_2$ with $Y_1 \subsetneq Y$, $Y_2 \subsetneq Y$ proper subsets.

Definition 2.2 An *affine variety* is an irreducible algebraic subset of the affine space, with the inherited Zariski topology.

Given a subset $X \subset \mathbb{A}_k^n$, define the *ideal* of X as

$$I(X) = \{f \in k[X_1, \ldots, X_n] : f(x) = 0 \text{ for all } x \in X\}.$$

By the Hilbert's Nullstellensatz (c.f. [6, 1. p. 85]), for any ideal $J \subset k[X_1, \ldots, X_n]$ we have

$$I(Z(J)) = \sqrt{J} := \left\{ f \in k[X_1, \ldots, X_n] : \text{there exists } r \geq 0 \text{ with } f^r \in J \right\}.$$

Then, there is a one-to-one correspondence between algebraic sets and radical ideals (which are ideals such that $J = \sqrt{J}$) and between affine varieties and prime ideals. Maximal ideals correspond to minimal algebraic sets, which are points in $\mathbb{A}_k^n$.

For any algebraic set $X = Z(\Sigma)$, the quotient ring

$$A(X) = k[X_1, \ldots, X_n]/I(X)$$

is called the *coordinate ring* of X. If X is an affine variety, $A(X)$ is a finitely generated k-algebra and an integral domain.

Examples of Affine Varieties

The zero locus of the polynomial $f(X_1, X_2) = X_1^2 + X_2^2 - 1 \in \mathbb{C}[X_1, X_2]$ is a conic X in the affine plane $\mathbb{A}_{\mathbb{C}}^2$, whose ideal is $I(X) = (f)$ which is a prime ideal, then X is an affine variety.

The zero locus of the polynomial $g(X_1, X_2) = X_1^2 X_2 \in \mathbb{C}[X_1, X_2]$ are the two axis in $\mathbb{A}_{\mathbb{C}}^2$, a subset which is not irreducible, hence $Z(S)$ is an algebraic set but not a variety. Indeed, we have

$$I(Z(S)) = I\left(Z(X_1^2 X_2)\right) = \sqrt{(X_1^2 X_2)} = (X_1 X_2),$$

which is a radical ideal but not a prime ideal.

Let us define the n-dimensional *projective space* over k, denoted by $\mathbb{P}_k^n$, as the quotient set of the vector space k^{n+1} minus the origin, by the equivalence relation

$$(x_0, x_1, \ldots, x_n) \sim (\lambda x_0, \lambda x_1, \ldots, \lambda x_n), \quad \lambda \in k\backslash\{0\}.$$

In the projective space, each point represents a line of k^{n+1}. Similarly we can define the projective space $\mathbb{P}(V)$ associated to a k-vector space V.

The ring of polynomials $R := k[X_0, X_1, \ldots, X_n]$ can be seen as a graded ring $R = \bigoplus_{d \geq 0} R_d$, decomposed as a direct sum of pieces given by homogeneous polynomials of each degree. Given a set of homogeneous polynomials $\Sigma \subset k[X_0, X_1, \ldots, X_n]$, let its zero set be given by

$$Z(\Sigma) = \left\{ x \in \mathbb{P}_k^n : f(x) = 0 \text{ for all } f \in \Sigma \right\}.$$

We call a subset $X \subset \mathbb{P}_k^n$ an *algebraic set* if there exists a set of homogeneous polynomials $\Sigma \subset k[X_0, X_1, \ldots, X_n]$ such that $X = Z(\Sigma)$. Similarly, we define the *Zariski topology* in the projective space by declaring the algebraic sets $Z(\Sigma)$ to be closed.

Definition 2.3 A *projective variety* is an irreducible algebraic subset of the projective space, with the inherited Zariski topology. A *quasi-projective variety* is a Zariski open subset of a projective variety. Note that affine varieties are quasi-projective.

For a subset $X \subset \mathbb{P}_k^n$, the *homogeneous ideal* of X, denoted by $I_h(X)$, is the ideal generated by all homogeneous polynomials vanishing at every point of X, where projective varieties correspond to homogeneous prime ideals not equal to $R_+ := \bigoplus_{d > 0} R_d$. And given an algebraic set $X = Z(\Sigma) \subset \mathbb{P}_k^n$, the quotient

$$A_h(X) = k[X_0, X_1, \ldots, X_n]/I_h(X)$$

is the *homogeneous coordinate ring* of X.

A projective variety X can be covered by affine varieties of the form $X \cap U_i$, where $U_i = \left\{ [x_0 : \ldots : x_n] \in \mathbb{P}_k^n : x_i \neq 0 \right\}$. Then, projective varieties can be understood as affine varieties glued together, the same way the projective space is.

Example of a Projective Variety

The zero locus of the homogeneous polynomial

$$f(X_0, X_1, X_2) = X_1^2 + X_2^2 - X_0^2 \in \mathbb{C}[X_0, X_1, X_2]$$

is an irreducible algebraic subset of $\mathbb{P}_\mathbb{C}^2$, then it is a projective variety which is the projectivization (i.e., by homogenizing the polynomial equations) of the affine conic seen in the previous affine example.

Given a projective variety $X \subset \mathbb{P}_k^n$, we define the *affine cone* of X as the affine variety $\hat{X} \subset k^{n+1}$ resulting of placing, for each $x \in X$, the corresponding line defining x in the projective space. The affine cone $\hat{X}$ intersects the origin and yields a cone such that its transverse slicings recover the projective variety X. The coordinate ring of the affine cone $\hat{X}$ coincides with the homogeneous coordinate ring of X, but seen on the affine space:

$$A(\hat{X}) = A_h(X) = k[X_0, X_1, \ldots, X_n]/I_h(X) \ .$$

Example of an Affine Cone

The affine cone of the variety $Z\left(X_1^2 + X_2^2 - X_0^2\right) \subset \mathbb{P}_k^2$ is the honest cone in $\mathbb{A}_k^3$ of the same equation $X_1^2 + X_2^2 - X_0^2 = 0$, which is singular at the origin. The slicing with planes $X_0 = c, 0 \neq c \in k$ gives the nondegenerate conic sections, which are affine varieties in $\mathbb{A}_k^3$ whose projectivization is X.

On an affine variety, we define *regular functions* as functions $f : X \to k$ such that for each $x \in X$, there is a neighborhood $x \in U \subset X$ where $f = g/h$ is described as a quotient of polynomials in $k[X_1, \ldots, X_n]$, and h does not vanish on U. This way, the coordinate ring recovers the ring of regular functions defined over an affine variety. If X is a projective algebraic variety, we require g and h to be homogeneous polynomials of the same degree for f to be regular. Given affine or projective varieties X and Y, a *morphism* between them $\alpha : X \longrightarrow Y$ is a continuous map verifying that for every Zariski open subset $U \subset Y$ and every regular function $f : U \longrightarrow k$, the composition $f \circ \alpha$ is also regular.

Grassmannians

Important examples of algebraic varieties are the *Grassmannians*. Given an n-dimensional k-vector space V, the set of all r-dimensional vector subspaces $W \subset V$ defines a projective variety $\mathrm{Gr}(r, n)$. Homogeneous coordinates can be given to this variety by the relations of the minors constructed with the generators of subspaces $W \subset V$, which are the Plücker coordinates embedding $\mathrm{Gr}(r, n)$ into the projective space. The Grassmannian $\mathrm{Gr}(1, n)$ actually corresponds to the projective space itself.

Complex Manifolds and Riemann Surfaces

Every algebraic variety X over $\mathbb{C}$ is a complex analytic manifold where each point $x \in X$ has a neighborhood biholomorphic to an open subset of $\mathbb{C}^n$. Moreover, if X is projective, X is a compact manifold.

In the case where X is a smooth complex projective variety of complex dimension 1, also called a *smooth complex projective curve*, X is a compact *Riemann surface* from the analytical point of view. Its *genus* is the genus of X as a topological compact surface and will be denoted by g.

2.1.2 Group Actions

Definition 2.4 A *linear or affine algebraic group* is an affine variety G with a group structure $(G, *)$ such that the multiplication map

$$* : G \times G \longrightarrow G, \quad (g, h) \longmapsto g * h$$

and the inverse map

$$(\cdot)^{-1} : G \longrightarrow G, \quad g \longmapsto g^{-1}$$

are morphisms of affine varieties.

If G is a smooth manifold and the multiplication and inverse maps are smooth, the structure above defines a *Lie group* G. If G is a complex manifold and the operations are analytic, we say that G is a *complex Lie group*. Complex affine algebraic groups will be the main examples of complex Lie groups.

A subgroup of an affine algebraic group G is an affine closed subvariety $H \subset G$ which is also a subgroup of G. Let

$$GL(n, \mathbb{C}) = \left\{ A \in M^{n \times n}(\mathbb{C}) : \det(A) \neq 0 \right\}$$

be the *complex general linear group*. It can be shown (c.f. [79, Theorem 1.1.3.3]) that every linear algebraic group is isomorphic to a subgroup of a general linear group, then all complex Lie groups can be seen as subgroups $G \subset GL(n, \mathbb{C})$.

Examples of Algebraic Groups

The main examples of algebraic groups come from distinguished matrix subgroups of the general linear group $GL(n, \mathbb{C})$. Examples are the *special linear group* $SL(n, \mathbb{C})$ of determinant 1 matrices, or the *symplectic group* $Sp(2n, \mathbb{C})$ of matrices preserving the standard symplectic form in $\mathbb{C}^{2n}$, which are matrices A verifying $A^t J A = J$ where

$$J = \begin{pmatrix} 0 & -I_n \\ I_n & 0 \end{pmatrix} \in M^{2n \times 2n}(\mathbb{C}) \ .$$

By complexifying real coordinates into complex ones, we can complexify real Lie groups K to obtain complex Lie groups $K_\mathbb{C}$. For example, $\mathbb{C}^* = GL(1, \mathbb{C})$ is the complexification of the unitary circle S^1, which is a real Lie group. In general, the group of unitary matrices $U(n)$ is a real Lie group which is a compact Lie subgroup of $GL(n, \mathbb{C})$, and its complexification is $U(n)_\mathbb{C} = GL(n, \mathbb{C})$. Another example is the group of real determinant 1 orthogonal matrices $SO(n, \mathbb{R})$ whose complexification is

$$SO(n, \mathbb{R})_\mathbb{C} = SO(n, \mathbb{C}) \subset GL(n, \mathbb{C}) \ ,$$

the complex special orthogonal group.

An (left) *action* of an algebraic group G on an affine or projective variety X, also called a *G-action* on X, is a morphism of varieties

$$\sigma : G \times X \longrightarrow X, \quad \sigma(g, x) = gx,$$

such that for every $g_1, g_2 \in G$, and $x \in X$:

$$g_1(g_2 x) = (g_1 \cdot g_2)x \quad \text{and} \quad ex = x,$$

where e is the identity element in G. The *stabilizer* of a point $x \in X$ is the closed subgroup of G given by

$$G_x := \{g \in G : gx = x\}$$

and the *orbit* of x under G is

$$O(x) := \{gx : g \in G\}.$$

A point $x \in X$ is an *invariant* point for the G-action if $gx = x$, for every $g \in G$. And, given a variety X with a G-action, a morphism of varieties $\alpha : X \longrightarrow Y$ is G-invariant if $\alpha(gx) = g\alpha(x) = \alpha(x)$, for every $g \in G, x \in X$, where we consider the trivial G-action on Y. Note that a G-invariant morphism is constant on each orbit.

Let X be an affine variety. A G-action on X defines an action on its coordinate ring $A(X)$ by

$$f \longmapsto f^g \in A(X), \quad \text{where} \quad f^g(x) := f(gx).$$

The *ring of invariants* is defined as

$$A(X)^G := \left\{ f \in A(X) : f^g = f \text{ for all } g \in G \right\}.$$

A *Lie algebra* is a k-vector space $\mathfrak{g}$ joint with a nonassociative, bilinear, alternating map, called *Lie bracket*:

$$[\cdot, \cdot] : \mathfrak{g} \times \mathfrak{g} \longrightarrow \mathfrak{g}, \quad (v, w) \mapsto [v, w],$$

satisfying Jacobi's identity

$$[u, [v, w]] + [v, [w, u]] + [w, [u, v]] = 0$$

for any $u, v, w \in \mathfrak{g}$. The main example is the Lie algebra associated to a Lie group, which we describe. Given G a real or complex Lie group, the tangent space at the identity $e \in G$ defines a Lie algebra $\mathfrak{g} := TG_e$ whose elements $X, Y \in TG_e$, by using the group law isomorphism, define smooth vector fields on G, and the Lie bracket is given by the matrix commutator $[X, Y] = XY - YX$ acting on smooth functions $f : G \to k$ as

$$[X, Y]f = (XY)f - (YX)f .$$

The relation between Lie groups G and Lie algebras $\mathfrak{g}$ is realized through the exponential map exp. Given $X \in \mathfrak{g}$ and a morphism $\gamma : \mathbb{R} \longrightarrow G$ such that $\gamma'(0) = X$, define $\exp(X) := \gamma(1)$. Unfortunately, this is not a one-to-one correspondence, because different Lie groups can have the same Lie algebra. Classifying Lie groups can thus be divided into classifying Lie algebras, whose classification is simpler, and classifying all group structures which have the same Lie algebra.

An important tool to classify Lie algebras is the *Killing form*

$$B(X, Y) = \text{tr} \, (\text{ad}(X) \circ \text{ad}(Y))$$

where $\text{ad}(X)(Y) = [X, Y]$, with $[\, , \,]$ the Lie bracket, is the *adjoint endomorphism*. It can be shown that, for simple algebras, the Killing form is the only possible bilinear symmetric adjoint-invariant form, up to rescaling.

Definition 2.5 Let V be an n-dimensional vector space over a field k. A *representation* of a Lie group G is a homomorphism

$$\xi : G \longrightarrow \text{GL}(V) := \text{GL}(n, k).$$

If there exists a proper subspace $W \subset V$ which is invariant for the representation $\xi \in \text{Hom} \, (G, \text{GL}(V))$, we say that ξ is *reducible* to $\text{GL}(W)$. Otherwise we say that ξ is an *irreducible* representation.

Definition 2.6 A complex Lie group is *reductive* if every representation splits into a direct sum of irreducible representations.

A special case of this is given by compact connected Lie groups K, because their complexifications $K_{\mathbb{C}}$ can be proved to be reductive groups. For example, $\text{GL}(n, \mathbb{C})$ and $\text{SL}(n, \mathbb{C})$ are reductive algebraic groups, complexifications of $\text{U}(n)$ and $\text{SU}(n)$ (determinant 1 unitary matrices), respectively.

Complex Algebraic Tori

An important type of complex algebraic Lie groups are *complex tori*, which are diagonalizable subgroups of $\text{GL}(n, \mathbb{C})$, then they can be described as a product of copies of the multiplicative group $\mathbb{C}^*$. Complex tori are the complexification of a product of n copies of the circle $S^1 = \text{U}(1)$; therefore, they are reductive.

2.2 Sheaf Theory and Schemes

To develop geometric invariant theory, we need to describe schemes, which are a generalization of algebraic varieties. Sheaves comprehend the abstraction of the ring of functions over a differential or algebraic variety. Their cohomology provides information about the geometry of the variety or the scheme themselves.

2.2.1 Sheaves and Cohomology

Definition 2.7 A *presheaf* $\mathcal{F}$ of abelian groups on a topological space X is the datum of an abelian group $\mathcal{F}(U)$ for each open subset $U \subset X$, and a restriction map $\varrho_{UV} : \mathcal{F}(U) \longrightarrow \mathcal{F}(V)$ for each inclusion $V \subset U$ of open sets, such that:

- $\mathcal{F}(\emptyset) = 0$
- $\varrho_{UU} = \mathrm{Id}_{\mathcal{F}(U)}$
- for inclusions $W \subset V \subset U$, we get the composition $\varrho_{VW} \circ \varrho_{UV} = \varrho_{UW}$.

A more concise definition is to say that a presheaf is a contravariant functor from the category of open sets of X and inclusions, to the category of abelian groups. In a similar way, we can define presheaves of rings or sets.

The elements of $\mathcal{F}(U)$ are called *sections* of $\mathcal{F}$ over U. Given a point $x \in X$, the *stalk* of $\mathcal{F}$ at x, denoted by $\mathcal{F}_x$, is the direct limit

$$\mathcal{F}_x := \varinjlim_{x \in U} \mathcal{F}(U) \, .$$

Homomorphisms of presheaves $\alpha : \mathcal{F} \to \mathcal{G}$ (over the same topological space X) are given by homomorphisms of abelian groups $\alpha|_U : \mathcal{F}(U) \to \mathcal{G}(U)$ for each open subset $U \subset X$, compatible with the restriction maps.

Definition 2.8 A presheaf $\mathcal{F}$ over X is a *sheaf* if it satisfies the additional conditions:

- (Uniqueness of sections) If two sections $s, t \in \mathcal{F}(U)$ coincide on each element of an open covering $U = \bigcup_{i \in I} U_i$, i.e., $s|_{U_i} = t|_{U_i}$ for all $i \in I$, then $s = t$.
- (Gluing) If over an open covering $U = \bigcup_{i \in I} U_i$ we have sections $s_i \in \mathcal{F}(U_i)$ compatible in the intersections, i.e., $s_i|_{U_i \cap U_j} = s_j|_{U_i \cap U_j}$ for all $i, j \in I$, there exists a unique $s \in \mathcal{F}(U)$ restricting to $s|_{U_i} = s_i$, for all $i \in I$.

Examples of sheaves

Given a smooth manifold X, $\mathcal{E}_X$ denotes the sheaf of smooth functions which assigns to each open subset $U \subset X$ the ring $\mathcal{E}_X(U)$ of smooth functions $f : U \to \mathbb{R}$. If X is a complex analytic manifold, O_X denotes the sheaf of rings of holomorphic functions. The stalk of each sheaf at a point $x \in X$ captures the ring of (smooth or holomorphic) germs at x. Smooth and holomorphic differential forms of degree k on X yield sheaves of rings denoted by $\mathcal{E}_X^k$ and Ω_X^k, respectively.

If X is an affine or projective variety over k, the rings of regular functions over each open subset $U \subset X$ form the sheaf of regular functions, also denoted by O_X. When $k = \mathbb{C}$, this sheaf coincides with the sheaf of holomorphic functions, hence the notation.

Given a sheaf of rings $\mathcal{A}$ over X, a *sheaf of modules* $\mathcal{F}$ over X is a sheaf of abelian groups such that, for each open subset $U \subset X$, $\mathcal{F}(U)$ is a module over the ring $\mathcal{A}(U)$. Sheaves of modules over algebraic varieties are usually sheaves of O_X-modules. The sheaf $\mathcal{F}$ is said to be *free* if $\mathcal{F}$ is isomorphic as a sheaf to $\mathcal{A} \times \cdots \times \mathcal{A}$, and *locally free* if the isomorphism holds locally for each $U \subset X$.

A subsheaf (resp. subpresheaf) $\mathcal{F}'$ of a sheaf (resp. presheaf) $\mathcal{F}$ is the choice of a subgroup $\mathcal{F}'(U) \subset \mathcal{F}(U)$ for each open subset, compatible with the restriction maps. A morphism of sheaves is a morphism of presheaves $\alpha : \mathcal{F} \to \mathcal{G}$ where $\mathcal{F}$ and $\mathcal{G}$ are sheaves.

If $\alpha : \mathcal{F} \to \mathcal{G}$ is a morphism of sheaves, we define the *sheaf kernel* $\mathrm{Ker}\,\alpha$ given by $\mathrm{Ker}\,\alpha(U) := \mathrm{Ker}\,[\alpha|_U : \mathcal{F}(U) \to \mathcal{G}(U)]$ on each open subset $U \subset X$. The analog for the image does not provide a sheaf; therefore, we define the *sheaf image* $\mathrm{Im}\,\alpha$ as the *sheaf associated to the presheaf* given by the image of each morphism $\alpha|_U : \mathcal{F}(U) \to \mathcal{G}(U)$. We associate a sheaf to a presheaf by adding all stalks and making it compatible with the restriction maps, in order to verify the sheaf conditions. Given a subsheaf $\mathcal{F}' \subset \mathcal{F}$, define the *quotient sheaf* $\mathcal{F}/\mathcal{F}'$ as the sheaf associated to the presheaf assigning the quotient $\mathcal{F}(U)/\mathcal{F}'(U)$ to each $U \subset X$. A morphism of sheaves is injective if $\mathrm{Ker}\,\alpha = 0$ and surjective if $\mathrm{Im}\,\alpha = \mathcal{G}$.

For a continuous map $f : X \to Y$ between topological spaces and a sheaf $\mathcal{F}$ on X, define the *push-forward sheaf* $f_*\mathcal{F}$ on Y by

$$f_*\mathcal{F}(V) := \mathcal{F}\left(f^{-1}(V)\right) ,$$

for each open subset $V \subset Y$. The *inverse image sheaf* of a sheaf $\mathcal{G}$ over Y, denoted by $f^{-1}(\mathcal{G})$, is the sheaf associated to the presheaf

$$f^{-1}(\mathcal{G})(U) := \varprojlim_{f(U) \subset V \subset Y} \mathcal{G}(V) .$$

Define the tensor product of two sheaves of O_X-modules $\mathcal{F}$ and $\mathcal{G}$, denoted by $\mathcal{F} \otimes_{O_X} \mathcal{G}$ (or $\mathcal{F} \otimes \mathcal{G}$ if the sheaf of rings is understood) as the sheaf associated to the presheaf with $\mathcal{F}(U) \otimes_{O_X} \mathcal{G}(U)$ for each open subset $U \subset X$. Given a continuous map $f : X \to Y$ between topological spaces and a sheaf of O_Y-modules $\mathcal{G}$, $f^{-1}(\mathcal{G})$ is a sheaf of $f^{-1}(O_Y)$-modules over X, and we define the *pull-back* of $\mathcal{G}$ by f to be

$$f^*\mathcal{G} := f^{-1}(\mathcal{G}) \otimes_{f^{-1}O_Y} O_X \ .$$

A sequence of sheaves and morphisms

$$\cdots \longrightarrow \mathcal{F}_{i-1} \xrightarrow{\alpha_{i-1}} \mathcal{F}_i \xrightarrow{\alpha_i} \mathcal{F}_{i+1} \longrightarrow \cdots$$

is said to be *exact* if $\operatorname{Ker} \alpha_i = \operatorname{Im} \alpha_{i-1}$, for each i; equivalently the stalks are equal $(\operatorname{Ker} \alpha_i)_x = (\operatorname{Im} \alpha_{i-1})_x$ for each $x \in X$. A *resolution* of a sheaf $\mathcal{F}$ is an exact sequence

$$0 \longrightarrow \mathcal{F} \longrightarrow \mathcal{F}_0 \longrightarrow \mathcal{F}_1 \longrightarrow \mathcal{F}_2 \longrightarrow \cdots \longrightarrow \mathcal{F}_i \longrightarrow \cdots \ .$$

Given a *short exact sequence* of sheaves

$$0 \longrightarrow \mathcal{F} \longrightarrow \mathcal{G} \longrightarrow \mathcal{H} \longrightarrow 0 \ ,$$

the induced sequence of global sections, i.e., sections on the total space X, is not necessarily exact on $\mathcal{H}(X)$:

$$0 \longrightarrow \mathcal{F}(X) \longrightarrow \mathcal{G}(X) \longrightarrow \mathcal{H}(X) \ .$$

The sheaf cohomology problem can be formulated as measuring the failure of this last sequence to be exact or, equivalently, lifting global sections of $\mathcal{H}$ to $\mathcal{G}$. For example, the sequence is exact for *soft* sheaves $\mathcal{F}$, which are those whose restrictions $\varrho_{XC} : \mathcal{F}(X) \to \mathcal{F}(C)$ to closed subsets $C \subset X$ are surjective.

Some sheaves, like the sheaf of smooth functions $\mathcal{E}_X$ or the sheaves of smooth differential k-forms $\mathcal{E}_X^k$ on a complex manifold X, are soft. Moreover, they are *fine* because they admit partitions of the unity: given an open covering $\mathcal{U} = \{U_i\}_{i=I}$ of X, there exists a collection of sheaf endomorphisms $\eta_i : \mathcal{F} \to \mathcal{F}$, supported on each U_i, summing up to the unity. However, the sheaf of holomorphic functions O_X is not soft.

Despite this, for any sheaf $\mathcal{F}$ it can be constructed (c.f. [92, II.3]) a canonical resolution by soft sheaves

$$0 \longrightarrow \mathcal{F} \longrightarrow \mathcal{F}_0 \longrightarrow \mathcal{F}_1 \longrightarrow \mathcal{F}_2 \longrightarrow \cdots \longrightarrow \mathcal{F}_i \longrightarrow \cdots$$

and using this resolution we can define sheaf cohomology.

Definition 2.9 The *p*th *cohomology group* of X with coefficients in a sheaf $\mathcal{F}$ is defined as

$$H^p(X, \mathcal{F}) = \frac{\text{Ker}\left(\mathcal{F}_p(X) \longrightarrow \mathcal{F}_{p+1}(X)\right)}{\text{Im}\left(\mathcal{F}_{p-1}(X) \longrightarrow \mathcal{F}_p(X)\right)}.$$

The 0th cohomology group $H^0(X, \mathcal{F})$ equals the global sections of the sheaf, $\mathcal{F}(X)$. In the special case when $\mathcal{F}$ is itself soft, it is also *acyclic* meaning that all cohomology groups $H^p(X, \mathcal{F}) = 0$, for $p > 0$.

De Rham Cohomology

Singular cohomology groups $H^p(X, \mathbb{R})$ of a smooth manifold X are computed by (sheaves of) groups of singular cochains, but it can also be computed via differential forms thanks to the De Rham theorem. It happens that there exists another resolution of the sheaf $\mathbb{R}_X$ of real locally constant functions by the sheaves of smooth differential k-forms,

$$0 \longrightarrow \mathbb{R}_X \longrightarrow \mathcal{E}_X^0 \longrightarrow \mathcal{E}_X^1 \longrightarrow \mathcal{E}_X^2 \longrightarrow \cdots \longrightarrow \mathcal{E}_X^k \longrightarrow \cdots,$$

then the sheaf cohomology groups can also be calculated as

$$H^p(X, \mathbb{R}) = H^p(X, \mathbb{R}_X) = \frac{\text{Ker}\left(\mathcal{E}_X^p(X) \longrightarrow \mathcal{E}_X^{p+1}(X)\right)}{\text{Im}\left(\mathcal{E}_X^{p-1}(X) \longrightarrow \mathcal{E}_X^p(X)\right)}.$$

A quite useful alternative way to compute sheaf cohomology is by using Čech *cohomology*. Let $\mathcal{U} = \{U_i\}_{i \in I}$ be an open covering of X and define a *p-simplex* σ as the tuple

$$\sigma\left(U_0, \ldots, U_p\right), \quad U_i \in \mathcal{U}, \ i = 0, \ldots, p,$$

with non-empty support $|\sigma| := \bigcap_{i=0}^{p} U_i \neq \emptyset$. Given a sheaf of abelian groups $\mathcal{F}$, the set of *p*-cochains $C^p(\mathcal{U}, \mathcal{F})$ is an abelian group whose elements are maps

$$f : \sigma \longmapsto f(\sigma) \in \mathcal{F}(|\sigma|).$$

The *coboundary* operator

$$\delta : C^p(\mathcal{U}, \mathcal{F}) \longrightarrow C^{p+1}(\mathcal{U}, \mathcal{F})$$

takes $f \in C^p(\mathcal{U}, \mathcal{F})$ to $\delta f \in C^{p+1}(\mathcal{U}, \mathcal{F})$ where, for $\sigma = (U_0, \ldots, U_{p+1})$, it is defined as

$$\delta f(\sigma) = \sum_{j=0}^{p+1} (-1)^j \varrho_{|\sigma_j||\sigma|}$$

with $\sigma_j = (U_0, \ldots, U_{j-1}, U_{j+1}, \ldots, U_{p+1})$. This coboundary operator satisfies $\delta^2 = 0$, then we get a *cochain complex*

$$C^0(\mathcal{U}, \mathcal{F}) \longrightarrow \cdots \longrightarrow C^{p-1}(\mathcal{U}, \mathcal{F}) \longrightarrow C^p(\mathcal{U}, \mathcal{F}) \longrightarrow C^{p+1}(\mathcal{U}, \mathcal{F}) \longrightarrow \cdots$$

whose cohomology is the Čech cohomology

$$\check{H}^p(\mathcal{U}, \mathcal{F}) := \frac{\mathrm{Ker}\left(C^p(\mathcal{U}, \mathcal{F}) \longrightarrow C^{p+1}(\mathcal{U}, \mathcal{F})\right)}{\mathrm{Im}\left(C^{p-1}(\mathcal{U}, \mathcal{F}) \longrightarrow C^p(\mathcal{U}, \mathcal{F})\right)} .$$

It happens that the direct limits of the Čech cohomology groups $\check{H}^p(\mathcal{U}, \mathcal{F})$ over all refinements of an open covering are isomorphic to the sheaf cohomology in Definition 2.9:

$$\check{H}^p(X, \mathcal{F}) := \varinjlim_{\mathcal{U}} \check{H}^p(\mathcal{U}, \mathcal{F}) \simeq H^p(X, \mathcal{F}) .$$

Moreover, if $\mathcal{U}$ is a *Leray covering* such that $\check{H}^p(|\sigma|, \mathcal{F}) = 0$, for all p-simplices $|\sigma|$ and $p \geq 1$, it is directly true that

$$\check{H}^p(\mathcal{U}, \mathcal{F}) \simeq H^p(X, \mathcal{F}) .$$

2.2.2 Schemes

For many applications, it is necessary to consider a broader notion than the one of algebraic variety. This leads to the idea of *scheme* which, roughly speaking, is an algebraic variety where we distinguish the nilpotency order. For example, the polynomials X and X^2 define the same variety in $\mathbb{A}_k^2$ (i.e., the Y-axis line), but the coordinate rings $k[X, Y]/(X)$ and $k[X, Y]/(X^2)$, i.e., the rings of regular functions defined over the varieties, are different. Affine schemes, together with their structure sheaf, will enlarge the category of algebraic varieties and regular functions.

Given a commutative ring R, the set of its prime ideals is its *spectrum* $\mathrm{Spec}(R)$. Let $I \subset R$ be an ideal of R, and denote by

$$V(I) = \{\mathfrak{p} \supset I : \mathfrak{p} \in \mathrm{Spec}(R)\} \subset \mathrm{Spec}(R)$$

the set of prime ideals containing I. By declaring the subsets $V(I)$ to be closed, we turn $\mathrm{Spec}(R)$ into a topological space. Note how this topology extends the Zariski topology for algebraic varieties by adding all prime ideals as extra points in $\mathrm{Spec}(R)$, not only the maximal ideals as we did for algebraic varieties.

Let $X = \mathrm{Spec}(R)$ be a topological space and endow it with a sheaf O_X whose sections encode the analog to the notion of regular functions on an algebraic variety. For each open subset $U \subset \mathrm{Spec}(R)$, define $O_X(U)$ to be the commutative ring with identity with sections

$$s : U \longrightarrow \bigsqcup_{p \in U} R_p$$

taking values in each localization R_p of R at the prime p, whose elements are quotients f/g of elements in R with $g \notin p$, then it is well defined at the point $p \in \mathrm{Spec}(R)$. Note that these sections $s \in O_X(U)$ take values in different local rings R_p, instead of taking values in the same field k, as it happens with varieties.

We define an *affine scheme* as the pair (X, O_X) where $X = \mathrm{Spec}(R)$ for some ring R and O_X is the previously defined sheaf of rings called the *structure sheaf* of X. The *stalk* $O_{X,p}$ of the structure sheaf at each point p is isomorphic to the local ring R_p which means that, locally, the germs of functions on each point p take values in the corresponding localization of the ring R, hence the name. A *scheme* is, in turn, a topological space X endowed with a sheaf of rings O_X where all stalks $O_{X,p}$ are local rings and, for every $x \in X$, there exists an open neighborhood $x \in U \subset X$ such that the restriction $(U, O_X|_U)$ is an affine scheme.

To pass from affine to projective schemes, we use the construction of the projective spectrum. Now let R be a graded ring and recall the notation $R_+ := \bigoplus_{d>0} R_d$. Remembering the correspondence between homogeneous ideals and projective varieties, $\mathrm{Proj}(R)$ will be the set of homogeneous ideals $q \in R$ not containing R_+. Similarly to the affine case, given a homogeneous ideal I, define the subsets

$$V(I) = \{q \supset I : q \in \mathrm{Proj}(R)\} \subset \mathrm{Proj}(R)$$

to be the closed subsets of $\mathrm{Proj}(R)$ as a topological space. Then, (X, O_X) defines a projective scheme where $X = \mathrm{Proj}(R)$ and the sections of the structure sheaf O_X, for each $U \subset \mathrm{Proj}(R)$, are defined as

$$s : U \longrightarrow \bigsqcup_{q \in U} R_{(q)}$$

taking values in the ring $R_{(q)}$ whose elements are fractions of homogeneous polynomials with denominator not contained in q; again, this way the value is well defined at the point $q \in \mathrm{Proj}(R)$. Note also how regular functions in a projective variety are in correspondence with elements of $R_{(q)}$ and, analogously, the stalks $O_{X,q}$ are isomorphic to $R_{(q)}$.

Let us finally mention the idea of *stack*. When taking quotients of varieties or schemes by the action of a group, which is precisely what GIT does and it appears in many constructions of moduli spaces, it generally happens that different points have different stabilizers. This prevents the whole quotient to be well defined as a scheme and we need to generalize to the category of stacks, to allow this phenomenon to occur. Although the formal definition of stack is far beyond the purpose of this manuscript, we will revisit the idea of quotients by groups and different stabilizers when talking about the notion of stability and, particularly, unstable objects.

2.3 Holomorphic Vector Bundles

One of the main tools in algebraic geometry are fibrations by vector spaces, i.e., *vector bundles*, to represent vector-valued functions on varieties. Here, we will introduce some basic definitions about vector bundles, and their geometric and topological properties.

2.3.1 Vector Bundles

Definition 2.10 A *smooth complex vector bundle* over a smooth manifold X is a smooth manifold E together with a smooth morphism $p : E \longrightarrow X$ with the following properties:

1. There is an open covering $X = \bigcup_{i \in I} U_i$ such that for every U_i there is a diffeomorphism h_i making the following diagram commutative

$$
\begin{array}{ccc}
E|_{U_i} & \xrightarrow{\ h_i\ } & U_i \times \mathbb{C}^r \\
{\scriptstyle p|_{U_i}} \downarrow & \swarrow {\scriptstyle p_1} & \\
U_i & &
\end{array}
$$

where p_1 is projection to the first factor (in other words, p is *locally trivial*).
2. For every pair (i, j), the composition $h_i \circ h_j^{-1}$ is linear on the *fibers* $E_x := p^{-1}(x)$, i.e.,

$$
h_i \circ h_j^{-1}(x, v) = \left(x, g_{ij}(v) \right),
$$

where $g_{ij} : U_i \cap U_j \longrightarrow \mathrm{GL}(r, \mathbb{C})$ is a smooth morphism.

The morphisms g_{ij} are called *transition functions*. For each $x \in X$, the fibers $E_x = p^{-1}(x)$ are finite dimensional vector spaces over $\mathbb{C}$ of the same dimension

$\dim(E_x) = r =: \mathrm{rk}(E)$, called the *rank* of E. The space E is called the *total space*, the continuous map $p : E \to X$ is the *projection map*, and X is the *base space*.

When X is a complex analytical manifold, the projection p and the transition functions g_{ij} are holomorphic, and the trivializing maps h_i are biholomorphic; this structure defines a *holomorphic vector bundle*.

> Frequently, the total space E denotes the vector bundle altogether, when the base space X and the projection p are clear from the context. In this text we will mostly refer to holomorphic vector bundles, omitting the word holomorphic where no confusion arises.

Definition 2.11 Given two vector bundles $p : E \to X$ and $q : F \to X$, a *vector bundle homomorphism* is a continuous map $\alpha : E \to F$ such that $q \circ \alpha = p$, and the restriction to fibers $\alpha|_{E_x} : E_x \to F_x$ is a linear map of vector spaces for each $x \in X$. The homomorphism α is an *isomorphism* if it is bijective and α^{-1} is continuous. We say then that E and F are *isomorphic*.

Vector Bundles Classified by Čech Cohomology

The set of isomorphism classes of vector bundles of rank r on X is canonically bijective to the Čech cohomology set $\check{H}^1\left(X, \underline{\mathrm{GL}(r, \mathbb{C})}\right)$, where $\underline{\mathrm{GL}(r, \mathbb{C})}$ is the locally constant sheaf of groups with this group on each open subset $U \subset X$. Choose a trivialization of a vector bundle E in open sets of a covering $\mathcal{U} = \{U_i\}_{i \in I}$ of X. The groups of p-cochains are given by

$$C^0\left(\mathcal{U}, \underline{\mathrm{GL}(r, \mathbb{C})}\right) = \{f_i : U_i \longrightarrow \mathrm{GL}(r, \mathbb{C})\}$$

$$C^1\left(\mathcal{U}, \underline{\mathrm{GL}(r, \mathbb{C})}\right) = \{g_{ij} : U_{ij} := U_i \cap U_j \longrightarrow \mathrm{GL}(r, \mathbb{C})\}$$

$$C^2\left(\mathcal{U}, \underline{\mathrm{GL}(r, \mathbb{C})}\right) = \{e_{ijk} : U_{ijk} := U_i \cap U_j \cap U_k \longrightarrow \mathrm{GL}(r, \mathbb{C})\}$$

and the coboundary maps are given by

$$C^0\left(\mathcal{U}, \underline{\mathrm{GL}(r, \mathbb{C})}\right) \xrightarrow{d^0} C^1\left(\mathcal{U}, \underline{\mathrm{GL}(r, \mathbb{C})}\right) \xrightarrow{d^1} C^2\left(\mathcal{U}, \underline{\mathrm{GL}(r, \mathbb{C})}\right)$$
$$f_i \longmapsto f_i \circ f_j^{-1}$$
$$g_{ij} \longmapsto g_{ij} \circ g_{kj}^{-1} \circ g_{ki}$$

We can identify each vector bundle E of rank r with an element of

$$\check{H}^1\left(X, \underline{\mathrm{GL}(r, \mathbb{C})}\right) = \check{H}^1\left(\mathcal{U}, \underline{\mathrm{GL}(r, \mathbb{C})}\right) = \frac{\mathrm{Ker}\, d^1}{\mathrm{Im}\, d^0},$$

by associating the element of $C^1\left(\mathcal{U}, \underline{GL(r, \mathbb{C})}\right)$ given by the transition functions g_{ij} on each intersection of the trivialization. This element is in $\operatorname{Ker} d^1$ because the functions g_{ij} satisfy the cocycle condition. Moreover, two elements associated to the same vector bundle differ by an element of $\operatorname{Im} d^0$, as the following diagram shows:

$$
\begin{array}{ccccc}
E|_{U_i} & \xrightarrow{\ \underset{\simeq}{h_i}\ } & U_i \times \mathbb{C}^r & \xrightarrow{\ f_i\ } & U_i \times \mathbb{C}^r \\
\| & & \downarrow{\scriptstyle g_{ij}} & & \downarrow{\scriptstyle f_j \circ g_{ij} \circ f_i^{-1}} \\
E|_{U_j} & \xrightarrow[\underset{\simeq}{h_j}]{} & U_j \times \mathbb{C}^r & \xrightarrow{\ f_j\ } & U_j \times \mathbb{C}^r
\end{array}
$$

Definition 2.12 Given an open subset $U \subset X$, a holomorphic *local section* of a vector bundle E on X is a holomorphic map $s : U \to E$ such that $(p \circ s)|_U = \operatorname{Id}|_U$. A holomorphic *global section* of a vector bundle E is a holomorphic map $s : X \to E$ such that $p \circ s = \operatorname{Id}|_X$

> The sections of a holomorphic vector bundle E over X form a locally free sheaf of modules over the structure sheaf O_X. We will denote this sheaf by the same letter E.

Then, we can use sheaf cohomology theory to study vector bundles. The cohomology groups of X with coefficients in a holomorphic vector bundle E have a natural structure of modules over $\mathbb{C}$, then $H^i(X, E)$ are complex vector spaces of dimension $h^i(X, E)$. In particular, the space of holomorphic global sections of a vector bundle E over X is $H^0(X, E)$ of dimension $h^0(X, E)$.

A vector bundle E is said to be *generated by global sections* if the evaluation morphism

$$
\begin{array}{rccc}
\mathrm{ev} : & H^0(X, E) \otimes O_X(U) & \twoheadrightarrow & E(U) \\
& s \otimes 1|_U & \longmapsto & s|_U
\end{array}
$$

is surjective for each open subset $U \subset X$. The morphism to be surjective means that

$$
(1, \dots, 1) \in O_X \oplus \cdots \oplus O_X \longmapsto s_1(x) + \cdots + s_n(x) \in E_x ,
$$

generate all the r-dimensional vector spaces E_x, for each $x \in X$.

The usual operations that we carry out on vector spaces and homomorphisms between them extend naturally to vector bundles, provided that those operations are performed fiberwise. For instance, let E, F be two vector bundles over X of ranks $\operatorname{rk}(E) = n$ and $\operatorname{rk}(F) = m$, respectively. We can define the following bundles:

- Direct sum $E \oplus F$, of rank $n + m$
- Tensor product $E \otimes F$, of rank $n \cdot m$
- Dual bundles $E^\vee$ and $F^\vee$, of ranks n and m
- Subbundles $F \subset E$, of rank m
- Quotient bundles E/F, of rank $n - m$
- Exterior powers $\wedge^k E$ for $k \leq n$, of rank $\binom{n}{k}$ and $\wedge^k F$ for $k \leq m$, of rank $\binom{m}{k}$
- Bundle of homomorphisms $\mathrm{Hom}(E, F)$, of rank $n \cdot m$

These bundles can be understood by means of operations in the transition functions g_{ij}. For example, transition functions of the direct sum $E \oplus F$ are block diagonal matrices given by

$$g_{ij}^E \oplus g_{ij}^F : U_i \cap U_j \longrightarrow \mathrm{GL}(n + m, \mathbb{C}) .$$

For a subset $F \subset E$ to define a subbundle, the dimension of the fibers $F_x = p^{-1}(x)$ must be constant for every $x \in X$. Similarly, not every subset defines a quotient E/F unless it is of constant rank fiberwise as well.

Tangent and Cotangent Bundle

Let X be a smooth complex projective curve, i.e., a compact Riemann surface. The *tangent bundle* TX is an example of a rank one vector bundle where the fibers E_x are just copies of $\mathbb{C}$ encoding all tangent vectors to X at the point x. The cotangent bundle, also called the *canonical line bundle* over X and denoted by $\omega_X := TX^\vee$, is the dual of the tangent bundle. Both are examples of *line bundles*, this is rank one vector bundles. In higher dimensional smooth projective varieties X, the canonical bundle is the line bundle defined as the highest exterior power of the cotangent bundle, $\omega_X := \wedge^n TX^\vee$, and coincides with the dualizing sheaf of next Serre's theorem.

Theorem 2.1 (Serre duality) *Given a vector bundle E over a smooth complex projective variety X of dimension n, there exist isomorphisms:*

$$H^i(X, E) \simeq H^{n-i}(X, E^\vee \otimes \omega_X), \quad i = 0, 1, \ldots, n.$$

In particular, when X is a smooth complex projective curve (i.e., of dimension $n = 1$):

$$H^1(X, E) \simeq H^0(X, E^\vee \otimes \omega_X).$$

2.3.2   Line Bundles

Line bundles are rank one vector bundles. They are also called *invertible sheaves* because a line bundle L has an inverse L^{-1} whose transition functions are the inverse functions $g_{ij}^{-1} : U_i \cap U_j \to \mathrm{GL}(1, \mathbb{C}) = \mathbb{C}^*$ and

$$L \otimes L^{-1} = L \otimes L^{\vee} = O_X.$$

The set of isomorphism classes of line bundles over X under the tensor product operation has a group structure called the *Picard group of X*, denoted by Pic X.

The structure sheaf O_X of a complex algebraic variety X is an example of a holomorphic line bundle. If X is projective, special interest has the line bundle $O_X(1)$ whose sections are restrictions to X of linear functions on the homogeneous coordinates in the projective space, called the *twisting sheaf of Serre*. Similarly, $O_X(n) := O_X(1)^{\otimes n}$ defines a line bundle whose sections are degree n polynomials. In the same vein, we denote by $E(n) := E \otimes O_X(n)$ the *twist* of E by n.

Definition 2.13 Let E be a vector bundle over X. Its *Euler characteristic*:

$$\chi(E) = \sum_{i \geq 0} (-1)^i h^i(X, E)$$

is the alternating sum of the dimensions of the cohomology spaces of E. We define the *Hilbert polynomial* of E as

$$P_E(m) = \chi(E(m)) ,$$

which turns out to be a polynomial on m of degree equal to the dimension of X.

Given a line bundle L on a variety X and a set of global sections $s_0, s_1, \ldots, s_n$ generating L under the evaluation map, with $\mathcal{L}\{s_0, s_1, \ldots, s_n\} = V \subset H^0(X, L)$, there exists a unique algebraic morphism into the projective space

$$\iota : X \longrightarrow \mathbb{P}_k^n \simeq \mathbb{P}(V^{\vee})$$

by associating each section with a coordinate of the ring $k[X_0, X_1, \ldots, X_n]$. The sections are pull-backs of the hyperplane coordinate functions $s_i = \iota^*(X_i)$, and the line bundle is the pull back of the twisting sheaf $L = \iota^*\left(O_{\mathbb{P}_k^n}(1)\right)$. This line bundle L is said to be *very ample* if ι is an immersion. The morphism being an immersion can be characterized in local terms by the condition that the subspace V of global sections generated by the $s_0, s_1, \ldots, s_n$ separates points (for two closed points $P, Q \in X$, there exists a section $t \in V$ vanishing at P but not at Q), and tangent vectors (for a closed point P and a tangent vector v, there exists a section $t \in V$ vanishing at P whose differential does not vanish at v). An *ample* line bundle is a line bundle L such that there is a power $L^{\otimes m}$ which is very ample. Often, we will tacitly assume that X is already embedded as a projective variety

$$X \hookrightarrow \mathbb{P}\left(H^0(X, L)^\vee\right)$$

by means of a very ample line bundle L identified with $\mathcal{O}_X(1)$.

Not every compact complex analytic manifold is projective in the sense that there exists an ample line bundle giving an embedding to a projective space. However, this is true for compact Riemann surfaces which have complex dimension one (c.f. [32, Section 2.1]) and hence, every compact Riemann surface is the same as a complex projective curve. These terms will be used interchangeably for the rest of the book.

Given an integer m, a vector bundle E over X is said to be *m-regular* if

$$H^i(X, E(m - i)) = 0, \text{ for every } i > 0 .$$

As a consequence of this definition, if a vector bundle E is m-regular then the twist $E(m)$ is generated by global sections, and also E is l-regular for all integers $l \geq m$.

Theorem 2.2 (Serre's vanishing theorem) *For each vector bundle E over X, there exists an m_0 such that E is m_0-regular and, for every $m \geq m_0$, the twisted bundle $E(m)$ is generated by global sections.*

2.3.3 Divisors

Let X be a compact Riemann surface. Define a *divisor D* in X as a finite formal sum of points:

$$D = \sum_i n_i x_i, \text{ where } n_i \in \mathbb{Z} \text{ and } x_i \in X.$$

The point-wise sum defines a sum for divisors, turning them into an abelian group $\mathrm{Div}(X)$. Define the *degree* of a divisor D as the sum of the coefficients

$$\deg(D) = \sum_i n_i .$$

A divisor is called *effective* if $n_i \geq 0$, for all $x_i \in X$.

For each $x \in X$, there exists a meromorphic function $h = f/g$, quotient of relatively prime holomorphic functions f and g on a neighborhood U of x, whose zeros are the points of $D \cap U$ with positive coefficients n_i and whose poles are those with negative coefficients. Let $X = \bigcup_{i \in I} U_i$ be an open covering of X where, for each U_i, D has meromorphic equation h_i. In $U_i \cap U_j$, the meromorphic function

$$g_{ij} := \frac{h_i}{h_j}$$

is a unitary element of $O_X^* \left(U_i \cap U_j \right)$ with no zeros or poles, where $O_X^* \left(U_i \cap U_j \right)$ denotes the maximal ideal in the ring $O_X \left(U_i \cap U_j \right)$. This defines functions

$$g_{ij} : U_i \cap U_j \longrightarrow \mathrm{GL}(1, \mathbb{C}) = \mathbb{C}^*$$

verifying the cocycle condition, therefore defines a line bundle denoted by $O_X(D)$.

Reciprocally, given a meromorphic section h of a line bundle L on X, h is described locally by a family of compatible quotients of polynomials $h_i = f_i / g_i$ on each $U_i \subset X$ such that

$$g_{ij} = \frac{h_i}{h_j} \in O_X^* \left(U_i \cap U_j \right) \ ,$$

hence the vanishing order $\mathrm{ord}_x(h_i)$ of h_i at a point $x \in U_i \cap U_j$, as a zero or pole, coincides with the vanishing order of h_j, $\mathrm{ord}_x(h_j)$. Then there is a globally well-defined vanishing order $\mathrm{ord}_x(h)$ for each $x \in X$ and h defines a divisor

$$D := \sum_i \mathrm{ord}_{x_i}(h) x_i \ ,$$

whose associated line bundle is $O_X(D)$. If the comparison functions are not units but actually one, i.e.,

$$g_{ij} = 1 \in O_X^* \left(U_i \cap U_j \right) \ ,$$

h is a global meromorphic function with $O_X(D) = O_X$ trivial, and the divisor D is called *principal*. Note that if h is holomorphic, the associated divisor is effective.

The order of the meromorphic sections h of the line bundle $O_X(D)$ at each point $x_i \in X$ verifies

$$\mathrm{ord}_{x_i}(h) \geq -n_i \ ,$$

where n_i is the coefficient of x_i in the definition of D. For example, if $D = -p$ is a single point divisor of degree -1, the meromorphic sections of the line bundle $O_X(D) = O_X(-p)$ necessarily vanish at p. Or, if $D = 2p + 2q$ of degree 4, the sections of $O_X(D) = O_X(2p + 2q)$ have, at most, poles of order 2 at p and q.

The group homomorphism

$$\mathrm{Div}(X) \longrightarrow \mathrm{Pic}(X) , \quad D + D' \longmapsto O_X(D) \otimes O_X(D')$$

has the subgroup of principal divisors as kernel, denoted by $\mathrm{Prin}(X) \subset \mathrm{Div}(X)$. Then we get the exact sequence

$$0 \longrightarrow \mathrm{Prin}(X) \longrightarrow \mathrm{Div}(X) \longrightarrow \mathrm{Pic}(X) \longrightarrow 0 .$$

Two divisors D and D' are linearly equivalent, denoted by $D \sim D'$, if their difference $D - D'$ is a principal divisor; therefore,

$$\mathrm{Pic}(X) \simeq \mathrm{Div}(X)/\!\sim\, \simeq \mathrm{Div}(X)/\mathrm{Prin}(X) .$$

The set of all effective divisors linearly equivalent to a given divisor D defines a *complete linear system*,

$$|D| := \mathbb{P}\left(H^0 (X, O_X(D)) \right)$$

whose elements are in correspondence with the holomorphic sections of the line bundle $O_X(D)$ defined by the divisor D. The dimension of $|D|$ equals $h^0 (X, O_X(D)) - 1$. Any projective subspace $V \subset |D|$ is a *linear system*. A linear system of low dimension is called a pencil, net or web if it is of dimension 1, 2, or 3, respectively.

If X is a compact Riemann surface, a global meromorphic function h on X has the same number of zeros and poles, counted with multiplicity, then principal divisors have degree zero. Therefore the group homomorphism degree

$$\deg : \mathrm{Div}(X) \longrightarrow \mathbb{Z}$$

descends to

$$\mathrm{Pic}(X) = \mathrm{Div}(X)/\mathrm{Prin}(X) \longrightarrow \mathbb{Z} ,$$

and we can define the degree of a line bundle L as the degree of the divisor given by any non-zero meromorphic section, which is the sum of zeros minus the sum of poles. For example, sections of the twisting sheaf of Serre $O_X(1)$ are linear polynomials given locally by equations $a X_0 + b X_1$, which vanish at a single point with no poles, hence $\deg (O_X(1)) = 1$. Using the group homomorphism between divisors and line bundles, we get $\deg (O_X(n)) = n$. And using the relationship between complete linear systems and spaces of global sections:

$$h^0 (X, O_X(n)) > 1 \quad \text{if and only if} \quad n > 0 ,$$

that way there are effective divisors of positive degree in the complete linear system.

Canonical Divisor

Denote by K the divisor associated to the canonical line bundle ω_X over a compact Riemann surface X. The degree of K and ω_X can be shown to be $2g - 2$, where g is the genus of X.

Combining the notion of linear system with the projective morphisms, we can understand each complete linear system defining a morphism to the projective space by means of its associated line bundle and its global sections. This morphism being a closed immersion can be characterized locally in terms of the elements of the linear system, whenever they separate points (i.e., for every two distinct points P and Q, there is a divisor passing through P and missing Q), and separate tangent vectors (i.e., for every point P and a tangent direction v at P, there is a divisor passing through P but with a different tangent at P).

The formal definition of the degree of a line bundle is given in terms of *Chern classes*, which are characteristic classes arising with holomorphic vector bundles. Chern classes for any vector bundle can be axiomatically defined from those of a line bundle L, whose only non-trivial class is the first one $c_1(L) \in H^2(X, \mathbb{Z})$. This, as a top cohomology class, coincides with the Euler class of the underlying smooth vector bundle. The first Chern class turns out to be a complete invariant for line bundles, because the morphism

$$c_1 : \operatorname{Pic} X \longrightarrow H^2(X, \mathbb{Z}) , \quad L \longmapsto c_1(L)$$

is a group homomorphism, by means of

$$c_1(L_1 \otimes L_2) = c_1(L_1) + c_1(L_2) ,$$

which is indeed an isomorphism.

From this, considering the *Chow ring* of a variety X consisting on algebraic cycles under the intersection product, we can extend the Chern classes to higher rank vector bundles by a set of axioms, particularly by the multiplicative property

$$c(E) = c(E') \cdot c(E'')$$

for exact sequences

$$0 \to E' \to E \to E'' \to 0 ,$$

where

$$c(E) = c_0(E) + c_1(E) + c_2(E) + \cdots + c_n(E)$$

is the *total Chern class* and each $c_i(E) \in H^{2i}(X, \mathbb{Z})$.

Let X be a smooth complex projective variety of dimension n, embedded in a projective space by means of an ample line bundle $O_X(1)$ corresponding to a divisor H, what is called a *polarization*. Given a vector bundle E, we define the *degree* of E by

$$\deg_H(E) = \int_X c_1(E) \wedge c_1(O_X(1))^{n-1} .$$

Note that, if X is a projective curve with $n = 1$, the degree is the integral of the first Chern class and does not depend on the polarization divisor H. This formal definition of degree agrees with the previous one in terms of divisors.

> If E is a vector bundle over X of rank r, we define the *determinant line bundle* as the top exterior product $\det(E) = \bigwedge^r E$. The degree of E equals the degree of its determinant line bundle $\det(E)$.

Once defined the degree, let us present some useful expressions to compute ranks and degrees of vector bundles. Recall that calculations of the ranks follow from counting dimensions of the corresponding vector spaces fiberwise. For the degree, note that sections of a degree d line bundle have d zeros minus poles counted with multiplicity; calculations follow by computing the number of zeros and poles from the corresponding section resulting of the bundle operation.

Then, for the direct sum:

$$\deg(E \oplus F) = \deg(E) + \deg(F) \quad \text{and} \quad \text{rk}(E \oplus F) = \text{rk}(E) + \text{rk}(F) .$$

For the tensor product:

$$\deg(E \otimes F) = \deg(E) \cdot \text{rk}(F) + \deg(F) \cdot \text{rk}(E)$$

$$\text{and} \quad \text{rk}(E \otimes F) = \text{rk}(E) \cdot \text{rk}(F) ,$$

which for line bundles L and M are:

$$\deg(L \otimes M) = \deg(L) + \deg(M) \quad \text{and} \quad \text{rk}(L \otimes M) = \text{rk}(L) = \text{rk}(M) = 1.$$

For duals:

$$\deg(E^\vee) = -\deg(E) \quad \text{and} \quad \mathrm{rk}(E^\vee) = \mathrm{rk}(E) .$$

For quotients:

$$\deg(E/F) = \deg(E) - \deg(F) \quad \text{and} \quad \mathrm{rk}(E/F) = \mathrm{rk}(E) - \mathrm{rk}(F) .$$

Besides, for the structure sheaf, $\deg(\mathcal{O}_X) = 0$ and, as we have seen before:

$$\deg(\mathcal{O}_X(1)) = 1 \quad \text{and} \quad \deg(\mathcal{O}_X(m)) = m .$$

As a final result, let us state the important Riemann-Roch theorem, relating the dimensions of the cohomology spaces of line bundles in a curve with their geometric invariants. When studying meromorphic functions, Riemann proved an important inequality between the dimension of the space of global holomorphic sections of a line bundle on a smooth complex projective curve (i.e., a compact Riemann surface), and the invariants of the bundle and the curve:

$$h^0(X, L) \geq \deg(L) + 1 - g .$$

Then his student, Gustav Roch, adds the *speciality term* completing an equality:

Theorem 2.3 (Riemann-Roch) *The Euler characteristic of a holomorphic line bundle L on a smooth complex projective curve X of genus g is*

$$\chi(L) = h^0(X, L) - h^1(X, L) = \deg(L) + 1 - g .$$

Applying Serre duality, $H^0\left(X, L^{-1} \otimes \omega_X\right) \simeq H^1(X, L)$, with ω_X the canonical line bundle, the Riemann-Roch theorem can be rephrased as

$$h^0(X, L) - h^0\left(X, L^{-1} \otimes \omega_X\right) = \deg(L) + 1 - g .$$

If E is a rank r vector bundle, Riemann-Roch theorem states that

$$\chi(E) = \deg(E) + r(1 - g) .$$

Chapter 3
Geometric Invariant Theory

Let G be a reductive complex algebraic group acting on an algebraic variety X. The purpose of Geometric Invariant Theory (abbreviated GIT) is to provide a way to define a quotient of X by the action of G with an algebro-geometric structure. Here we present a sketch of the treatment; for a deeper understanding and proofs see the seminal work [66] (and the extended version [67]) and the references [65, 79].

In the first sections we review the idea of quotient of X by G, the issues arising, and how GIT provides an adequate solution to the quotient problem. This solution relies on the idea of stability which is checked by 1-dimensional subgroups of G, as the Hilbert–Mumford criterion states. After this, we revisit the notion of stability from the differential and symplectic points of view.

Examples of different quotient problems are provided to understand how these notions work. In the final section, we explore the concept of maximal unstability which establishes a hierarchy among the unstable points, the ones removed when taking the GIT quotient.

3.1 Quotients and the Notion of Stability

The problem of taking the quotient of a variety X by the action of an algebraic group G has been extensively studied in mathematics. In the case when the variety X is affine there is a simpler solution which dates back to Hilbert's 14th problem. Let $A(X)$ be the coordinate ring of the affine variety X. Nagata [69] proved that, if the algebraic group G is reductive, the ring of invariants $A(X)^G$ is finitely generated; hence, it is the coordinate ring of an affine variety, and therefore we can define the quotient X/G as the affine variety associated to the ring $A(X)^G$.

Fig. 3.1 Orbits of the action
in Example 3.1

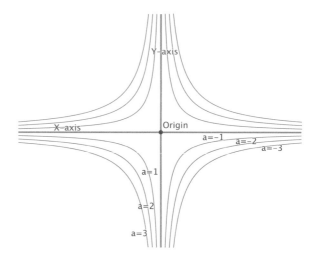

Fig. 3.1 Orbits of the action
in Example 3.1

When taking the quotient of an algebraic variety by an algebraic group G, there
are some issues which have to be taken into account. First one has to do with the
separateness of the quotient space and will led us to the identification of orbits under
the action of G. Here it is a simple example which shows this feature in the affine
case.

Problem: Non-separated orbits in a quotient

Example 3.1 Consider the action

$$\sigma :\ \mathbb{C}^* \times \mathbb{C}^2 \longrightarrow \mathbb{C}^2$$
$$(\lambda, (x, y)) \longmapsto (\lambda x, \lambda^{-1} y)$$

where real sections of the orbits are represented in Fig. 3.1. These orbits are the
hyperboles $xy = a$, with a as constant, plus three special orbits, the x-axis, the
y-axis, and the origin. Observe that the origin lies in the closure of the x-axis
and the y-axis. Note that all hyperboles are separated, even infinitesimally, by
the homogeneous polynomial XY: the differential of the function XY along the
transverse direction of the orbits is non zero, whereas for the three special orbits
(the two axes and the origin), while they are separated from the other orbits by the
polynomial XY, none of them is infinitesimally separated from the other two.

The coordinate ring of $\mathbb{C}^2$ is $\mathbb{C}[X, Y]$ and the ring of invariants is

$$\mathbb{C}[X, Y]^{\mathbb{C}^*} \simeq \mathbb{C}[XY] \simeq \mathbb{C}[Z] ,$$

then it does not distinguish between the three special orbits and identifies them in a
unique single point in the quotient space. Hence, the orbit space (the space where

each point corresponds to an orbit) is non-separated, but the quotient space is the variety associated to the ring of invariants $\mathbb{C}[X, Y]^{\mathbb{C}^*} \simeq \mathbb{C}[Z]$, that is, the affine line $\mathbb{A}_{\mathbb{C}}^1 = A(\mathbb{C}[Z])$ which is separated.

Once we know how to take quotients of affine varieties, let us deal with the projective case. We can guess that, as projective varieties are given by gluing affine pieces, we can take the quotient of each affine piece and then paste them together. As we want these pieces to be respected by the action of G, we want them to be G-invariant, hence we are looking for subsets of the form

$$X_f = \{x \in X : f(x) \neq 0\}$$

which are G-invariant or, equivalently, looking for G-invariant functions f. First note, as the following example shows, that the action of G on a projective variety X does not determine an action on the graded ring $\mathbb{C}[X_0, \ldots, X_n]$, or a quotient of it.

Problem: G-action on a projective variety X does not determine an action on the graded ring

Let $\mathbb{C}^*$ act on $\mathbb{P}_{\mathbb{C}}^1$ trivially, that is:

$$\lambda \cdot [x_0 : x_1] = [\lambda x_0 : \lambda x_1] = [x_0 : x_1], \quad \text{for} \quad \lambda \in \mathbb{C}^* .$$

This action is compatible with the trivial action of $\mathbb{C}^*$ on $\mathbb{C}[X_0, X_1]$ which acts as $\lambda \cdot f(X_0, X_1) = f(X_0, X_1)$, but it is also compatible with the action $\lambda \cdot f(X_0, X_1) = (\lambda f)(X_0, X_1)$, which multiplies each homogeneous polynomial by the corresponding scalar.

Consider X embedded in a projective space $X \hookrightarrow \mathbb{P}_{\mathbb{C}}^n$ by a very ample line bundle $\mathcal{O}_X(1)$, where each homogeneous coordinate comes from evaluating a linear section $s \in H^0(\mathcal{O}_X(1))$. The action of G lifts to an action on the total space of $\mathcal{O}_X(1)$ and acts on each $H^0(\mathcal{O}_X(m))$, where we can see a section $s \in H^0(\mathcal{O}_X(m))$ as a homogeneous polynomial of degree m. We can identify $\bigoplus_{m \geq 0} H^0(X, \mathcal{O}_X(m))$ with the coordinate ring of the affine cone $A(\hat{X})$; therefore, this is equivalent to linearizing the action to the affine cone $\hat{X} \subset \mathbb{C}^{n+1}$, meaning to give an action on $\mathbb{C}^{n+1}$ being the former action of G when restricted to X. Let

$$\bigoplus_{m \geq 0} H^0(X, \mathcal{O}_X(m))^G$$

be the invariant graded ring formed by invariant pieces of each degree, which is finitely generated (c.f. [89, Lemma 3.3]). We define the quotient of X by G, in the projective case, by

$$\text{Proj} \bigoplus_{m \geq 0} H^0(X, O_X(m))^G = \text{Proj}\, A(\hat{X})^G \,.$$

Different powers of the linearization $O_X(m)$, $m > 0$, give different embeddings into the projective space, and we want to understand the points of the quotient of X by G under the morphisms:

$$X \dashrightarrow \mathbb{P}\left(\left(H^0(X, O_X(m))^G\right)^{\vee}\right), \quad x \longmapsto s(x),$$

whose images are the evaluation of the G-invariant sections $s \in H^0(O_X(m))$. Then, the points in the quotient are defined in the complement of the vanishing locus of G-invariant polynomials of degree ≥ 1; therefore, the points where all G-invariant polynomials of degree ≥ 1 vanish cannot appear in the quotient. This motivates the following:

Definition 3.1 A point $x \in X$ is called *GIT semistable* if there exists a G-invariant section $s \in H^0(X, O_X(m))^G$, $m \geq 1$, such that $s(x) \neq 0$. If, moreover, the orbit of x is closed, it is called *GIT polystable*, and if, furthermore, this closed orbit has the same dimension as G (i.e. if x has finite stabilizer), then x is called a *GIT stable* point. We say that a closed point $x \in X$ is *GIT unstable* if it is not GIT semistable.

This way, the notion of GIT stability depends on the embedding and the linearization, that is, it depends on a line bundle and a lifting of the action to the total space of this line bundle.

> In the case where X is affine, there are no restrictions for the points to be in the projective quotient. The constants are always G-invariant functions and all points can be considered semistable. We recover $\text{Spec}\left(A(X)^G\right)$ as the quotient of X by G.

Mumford [66] developed its Geometric Invariant Theory to give a meaningful geometric structure to the quotient of X by G. It turns out that, for the semistable orbits, we can give a pleasant solution to our quotient problem.

Definition 3.2 Let G be an algebraic group acting on a scheme X. A *good quotient* is a scheme Y with a G-invariant morphism $p : X \to Y$ such that

1. p is surjective and affine (for every $V \subset Y$ affine, $p^{-1}(V) \subset X$ is affine).
2. $p_*\left(O_X^G\right) = O_Y$, where O_X^G is the sheaf of G-invariant functions on X.

3. If Z is a closed G-invariant subset of X, then $p(Z)$ is closed in Y. Furthermore, if Z_1 and Z_2 are two closed G-invariant subsets of X with $Z_1 \cap Z_2 = \emptyset$, then $p(Z_1) \cap p(Z_2) = \emptyset$.

Theorem 3.1 ([66, Proposition 1.9, Theorem 1.10]) *Let G be a reductive algebraic group acting on a projective scheme X. Let X^{ss} (resp. X^s) be the subset of GIT semistable points (resp. GIT stable). Both X^{ss} and X^s are open subsets. There exists a good quotient $p : X^{ss} \to Y$ where closed points are in one-to-one correspondence to the orbits of GIT polystable points, Y is a projective scheme, the image $Y^s := p(X^s)$ is open, and the restriction $p|_{X^s} : X^s \to Y^s$ is a* geometric quotient *where each point corresponds to one orbit.*

S-equivalence

Two orbits whose closures have non-empty intersection will be called S-*equivalent* and will define the same point in the quotient. GIT proves that there is only one closed orbit within each equivalence class (the orbit which is called polystable). The points of the moduli space are in correspondence with these distinguished closed orbits, therefore we can say that the moduli that we obtain classifies polystable points or S-*equivalence* classes of points. The S in the notion of S-equivalence comes from Seshadri [80, 81].

We will denote the good quotient of Theorem 3.1 by $X^{ss} \to X^{ss} /\!\!/ G$. The use of double slash $/\!\!/$ means that we make two identifications: one is the identification of the points of each orbit; the other one is the identification of S-equivalent orbits. We will refer to $X^{ss} /\!\!/ G$ as the *GIT quotient* of X by G. The geometric quotient will be denoted by $X^s \to X^s / G$ where we remove the second slash because, by definition, each point in a geometric quotient corresponds to a single orbit, without a second further identification.

Note that a GIT quotient provides also a line bundle in the quotient

$$\iota : X^{ss} /\!\!/ G \hookrightarrow \mathbb{P}\left(\left(H^0\left(X, O_X(m)\right)^G \right)^{\vee} \right) := \mathbb{P}^n_k$$

defined as $\iota^*\left(O_{\mathbb{P}^n_k}(1) \right)$.

Next we start to analyze the first example: the construction of the projective space as a GIT quotient.

Projective space as a GIT quotient

Example 3.2 Let $\mathbb{C}^* \times \mathbb{C}^{n+1} \longrightarrow \mathbb{C}^{n+1}$ be the scalar action, that is,

$$\lambda \cdot (z_0, z_1, \ldots, z_n) = (\lambda z_0, \lambda z_1, \ldots, \lambda z_n) , \quad \lambda \in \mathbb{C}^*, \quad (z_0, z_1, \ldots, z_n) \in \mathbb{C}^{n+1} .$$

If we consider the affine quotient, the only invariant functions are the constants, then:

$$A\left(\mathbb{C}^{n+1}\right)^{\mathbb{C}^*} = \mathbb{C}[X_0, X_1, \ldots, X_n]^{\mathbb{C}^*} = \mathbb{C}$$

and hence the affine quotient will be a single point: $\mathrm{Spec}(\mathbb{C})$.

To have more invariant functions and, then, a richer quotient space when applying GIT, we will consider invariant sections of the lifted action by characters, called *semi-invariants* (see [65, Section 6.b]). For a complex group G, f is a G-semi-invariant function with respect to a character $\chi : G \to \mathbb{C}^*$ if

$$f(g \cdot x) = \chi(g) \cdot f(x) .$$

If G acts on an affine variety X, the set of semi-invariants for the character χ is denoted by $A(X)_\chi^G$, and $\bigoplus_{m \geq 0} A(X)_{\chi^m}^G$ is a finitely generated graded ring. We define the GIT quotient as

$$X_\chi^{ss} /\!\!/ G := \mathrm{Proj} \bigoplus_{m \geq 0} A(X)_{\chi^m}^G ,$$

where a point x is semistable with respect to χ if there exists a G-semi-invariant section s with respect to χ^m, $m \geq 1$, such that $s(x) \neq 0$.

Define, for each $p \in \mathbb{Z}$, the characters

$$\chi_p : \mathbb{C}^* \longrightarrow \mathbb{C}^* , \quad \lambda \longmapsto \lambda^p .$$

The $\mathbb{C}^*$-semi-invariants will be those sections which verify

$$s(\lambda(z_0, z_1, \ldots, z_n)) = \chi_p^m(\lambda) \cdot s(z_0, z_1, \ldots, z_n) = \lambda^{pm} \cdot s(z_0, z_1, \ldots, z_n) .$$

- If $p > 0$, the $\mathbb{C}^*$-semi-invariant sections are given by homogeneous polynomials of degree pm. Then, all points are semistable except the origin, and, moreover, they are closed orbits of maximal dimension; hence, $X^{ss} = X^s = \mathbb{C}^{n+1} \backslash \{0\}$ and the quotient is the projective space denoted by

$$X_{\chi_p}^{ss} /\!/ \mathbb{C}^* = \mathrm{Proj} \bigoplus_{m \geq 0} \mathbb{C}[X_0, X_1, \ldots, X_n]_{pm} =: \mathbb{P}_{\mathbb{C}}^n .$$

Note that, even though the GIT quotient space is the same for every p, the line bundles of the quotient are different: $O_{\mathbb{P}_{\mathbb{C}}^n}(p)$, for each $p > 0$.

- If $p < 0$, note that there are no $\mathbb{C}^*$-semi-invariant sections; hence, all points are unstable and the quotient is empty.
- If $p = 0$, the character is trivial and the $\mathbb{C}^*$-semi-invariant sections are the same as the invariant functions in the affine quotient, which are the constants. Then, all points are semistable and the projective spectrum turns out to be

$$X_{\chi_0}^{ss} /\!/ \mathbb{C}^* = \mathrm{Proj} \bigoplus_{m \geq 0} \mathbb{C}[X_0, X_1, \ldots X_n]_0 = \mathrm{Proj}\, \mathbb{C}[Z] = \mathrm{Spec}(\mathbb{C}) = \mathrm{Spec}\, A(X)^G$$

which is the single point of the affine quotient.

This example shows how GIT stability of the orbits depends, in an essential way, on the different choices of linearization, giving completely different GIT quotients for different linearizations.

3.2 Hilbert–Mumford Criterion

To determine whether an orbit is GIT stable or unstable, we have to calculate invariant functions. This calculation is quite involved and dates back to Hilbert. One of Mumford's major achievements was to give a very simple numerical criterion to determine GIT stability, called in the literature the *Hilbert–Mumford criterion*.

Let X be a projective variety with an action of a reductive algebraic group G. It can be proved that a point x is GIT semistable if and only if $0 \notin \overline{G \cdot \hat{x}}$, where $\hat{x}$ lies over x in the affine cone $\hat{X}$. Intuitively one direction is clear: x is a GIT unstable point if, for all $m \geq 1$, all G-invariant homogeneous polynomials vanish at x; as all homogeneous polynomials (in particular the G-invariant ones) vanish at zero, the points containing zero in the closure of their orbits will be GIT unstable by continuity. The converse can be seen in [71, Proposition 4.7] or [66, Proposition 2.2].

The essence of the Hilbert–Mumford criterion is that GIT stability for the whole group G can be checked through 1-*parameter subgroups*

$$\rho : \mathbb{C}^* \longrightarrow G$$

such that we can reach every point in the closure of an orbit through them. Hence, a point is GIT semistable for the action of G if and only if it is so for the action

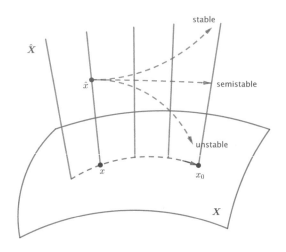

Fig. 3.2 GIT stability of orbits through 1-parameter subgroups. The different limits in the affine cone $\hat{X}$ of a projective variety X yield different stability denominations

of every 1-parameter subgroup. Then, together with the observation of the previous paragraph, GIT stability measures whether 0 belongs to the closure of the lifted orbit or not, a belonging which can be checked through 1-dimensional paths (Fig. 3.2).

Theorem 3.2 *Let $\hat{x}$ be a point in the affine cone $\hat{X}$ over X, lying over $x \in X$.*

- *x is semistable if, for all 1-parameter subgroups ρ, there exists $\lim\limits_{t \to 0} \rho(t) \cdot \hat{x} \neq 0$ or $\lim\limits_{t \to 0} \rho(t) \cdot \hat{x} = \infty$.*
- *x is polystable if it is semistable and the orbit of $\hat{x}$ is closed.*
- *x is stable if, for all non-trivial 1-parameter subgroups ρ, $\lim\limits_{t \to 0} \rho(t) \cdot \hat{x} = \infty$ (then the stabilizer of x is finite).*
- *x is unstable if there exists a 1-parameter subgroup ρ such that $\lim\limits_{t \to 0} \rho(t) \cdot \hat{x} = 0$.*

To apply the previous Hilbert–Mumford criterion to a quotient of an affine variety by the action of a reductive group linearized by means of a character, as in Example 3.2, we do the following variation (see [58, Section 2]). Let $x \in X$ and a lift of x to the affine cone $\hat{x} = (x, w) \in X \times (k\backslash\{0\})$. Given a character $\chi : G \to \mathbb{C}^*$, the lifting of the action to the affine cone is defined as $g \cdot (x, w) = (gx, \chi^{-1}(g)w)$. A point is semistable if $\overline{G \cdot \hat{x}} \cap (X \times \{0\}) = \emptyset$ and stable if $\overline{G \cdot \hat{x}}$ is closed and the stabilizer of $\hat{x}$ is finite. Equivalently, x is semistable if for all 1-parameter subgroups ρ, $\lim\limits_{t \to 0} \rho(t) \cdot \hat{x} \notin X \times \{0\}$, and stable if the only 1-parameter subgroup for which that limit exists is the trivial one.

Stability in the projective space as a GIT quotient

Example 3.3 (Continuation of Example 3.2) Given $z := (z_0, z_1, \ldots, z_n) \in \mathbb{C}^{n+1}$, let $\hat{z} := (z, w) \in \mathbb{C}^{n+1} \times (\mathbb{C}\setminus\{0\})$ be a point in the affine cone lying over z. Let $\chi_p(\lambda) = \lambda^p$ be a character and let

$$\rho_k : \mathbb{C}^* \longrightarrow \mathbb{C}^*, \quad t \longmapsto t^k$$

be the possible 1-parameter subgroups of $\mathbb{C}^*$ for each $k \in \mathbb{Z}$, where note that $k = 0$ gives the trivial one. The lifted action to the affine cone is given by

$$\lambda \cdot (z, w) = (\lambda z, \chi^{-1}(\lambda)w) = (\lambda z, \lambda^{-p}w)$$

and the limits of the 1-parameter subgroups are, for each $k \in \mathbb{Z}$:

$$\lim_{t \to 0} \rho_k(t) \cdot \hat{z} = \lim_{t \to 0} \rho_k(t) \cdot (z, w) = \lim_{t \to 0} (t^k z, \chi_p^{-1}(\rho_k(t))w) = \lim_{t \to 0} (t^k z, t^{-kp}w) .$$

- If $p > 0$, for a point $z \neq 0$ and every k we have $\lim_{t \to 0} \rho_k(t) \cdot \hat{z} \notin \mathbb{C}^{n+1} \times \{0\}$, then all $z \neq 0$ are semistable. Moreover, the only k for which the limit exists is $k = 0$, then all $z \neq 0$ are stable. However, a lift of $z = 0$ is $\hat{z} = (0, w)$ and, if $k < 0$, $\lim_{t \to 0} (0, t^{-kp}w) = (0, 0)$, then the origin is unstable.
- If $p < 0$, for every point z and $k > 0$ we have $\lim_{t \to 0} \rho_k(t) \cdot \hat{z} = (0, 0) \in \mathbb{C}^{n+1} \times \{0\}$, then all points are unstable.
- If $p = 0$, for all z and all k we have $\lim_{t \to 0} \rho_k(t) \cdot \hat{z} = \lim_{t \to 0} (t^k z, w) \notin \mathbb{C}^{n+1} \times \{0\}$, then all points are semistable. The limit always exists for all $k > 0$, then all points are semistable but not stable.

Given ρ a 1-parameter subgroup of G and given $x \in X$ we can define

$$P_x : \mathbb{C}^* \longrightarrow X, \quad P_x(t) = \rho(t) \cdot x .$$

The limit

$$x_0 := \lim_{t \to 0} \rho(t) \cdot x$$

(which always exists because X is projective) is clearly a fixed point of the action of $\mathbb{C}^*$ on X induced by ρ. Thus, $\mathbb{C}^*$ acts on the fiber of the line bundle over x_0, say, with weight ρ_x. One defines the numerical function

$$\vartheta(x, \rho) := \rho_x$$

called the *weight* of the action of the 1-parameter subgroup ρ over x.

Having defined the weight of the action $\vartheta(x, \rho)$, we are ready to state the Hilbert–Mumford numerical criterion of GIT stability:

Theorem 3.3 (Hilbert–Mumford Numerical Criterion) *[66, Theorem 2.1], [71, Theorem 4.9] With the previous notations:*

- x is semistable *if for all 1-parameter subgroups ρ, $\vartheta(x, \rho) \leq 0$.*
- x is polystable *if x is semistable and for all 1-parameter subgroups ρ such that $\vartheta(x, \rho) = 0$, there exists an element $g \in G$ with $x_0 = g \cdot x$.*
- x is stable *if for all nontrivial 1-parameter subgroups ρ, $\vartheta(x, \rho) < 0$.*
- x is unstable *if there exists a 1-parameter subgroup ρ such that $\vartheta(x, \rho) > 0$.*

The 1-parameter subgroups induce a linear action of $\mathbb{C}^*$ on the total space of the line bundle, which we think locally as $\mathbb{C}^{n+1}$ for an n-dimensional projective variety X. By a result of Borel (c.f. [10, section 4.6]), an action like that can be diagonalized such that, in coordinates of a basis of $\mathbb{C}^{n+1}$, the action can be expressed as

$$\rho(t) \cdot \hat{x} = \rho(t) \cdot \left(\hat{x}_0, \hat{x}_1, \ldots, \hat{x}_n \right) = \left(t^{\rho_0} \hat{x}_0, t^{\rho_1} \hat{x}_1, \ldots, t^{\rho_n} \hat{x}_n \right) .$$

Taking into account this, the definition of the weight $\vartheta(x, \rho)$ can be restated as

$$\vartheta(x, \rho) = \min \left\{ \rho_i : \hat{x}_i \neq 0 \right\} .$$

To understand this notion we can observe that, in the diagonal basis for the 1-parameter subgroup, the limit of the action can be taken coordinate-by-coordinate. If all exponents ρ_i are positive, the minimum is positive and $\lim_{t \to 0} \rho(t) \cdot \hat{x} = 0$, which corresponds to an unstable point. Conversely, if $\rho_i = 0$, the limit of the coordinate $\hat{x}_i$ is the same $\hat{x}_i$, and, if $\rho_i < 0$, the limit goes to infinity: hence if the minimum of the exponents is ≤ 0, the point is semistable, and, if it is strictly negative, it is stable.

In the case of affine quotients by an action linearized with a character, as in Examples 3.2 and 3.3, following [58, Section 2] the weight is computed by means of the pairing between characters χ and 1-parameter subgroups ρ, such that $\langle \chi, \rho \rangle = m$ if $\chi(\rho(t)) = t^m$. Then, x is semistable if for all 1-parameter subgroups ρ for which the limit $\lim_{t \to 0} \rho(t) \cdot x := x_0$ exists, we have $\langle \chi, \rho \rangle \geq 0$; and x is stable if the only 1-parameter subgroup ρ for which the limit exists and also verifies $\langle \chi, \rho \rangle = 0$ is the trivial one. Note that the change of sign is due to the inverse of the character in the lifted action, as we see in the following example.

Hilbert–Mumford criterion in the projective space as a GIT quotient

Example 3.4 (Continuation of Examples 3.2 and 3.3) Given a 1-parameter subgroup

$$\rho_k : \mathbb{C}^* \longrightarrow \mathbb{C}^* , \quad t \longmapsto t^k ,$$

the pairing with the character $\chi_p(\lambda) = \lambda^p$ is $\langle \chi, \rho \rangle = kp$. The limit on a point $z \in \mathbb{C}^{n+1}$ is given by

$$\lim_{t \to 0} \rho_k(t) \cdot z = \lim_{t \to 0} t^k z .$$

If $z \neq 0$, this limit exists if and only if $k \geq 0$, while if $z = 0$, it exists for every k.

- If $p > 0$, points $z \neq 0$ are semistable because the limit exists for all $k \geq 0$, with $\langle \chi, \rho \rangle = kp \geq 0$. And they are also stable because the only k for which $\langle \chi, \rho \rangle = 0$ is $k = 0$. The origin $z = 0$ is unstable because the limit also exists for $k < 0$, with $\langle \chi, \rho \rangle = kp < 0$.
- If $p < 0$, all points z are unstable because for $k > 0$ the limit exists with $\langle \chi, \rho \rangle = kp < 0$.
- If $p = 0$, any $k > 0$ gives a nontrivial 1-parameter subgroup for which the limit exists for every z with $\langle \chi, \rho \rangle = kp = 0$, then all points are semistable but not stable.

The next example is the fundamental one: the moduli space of binary forms or configurations of n points in the projective line. This example is originally due to Hilbert and it is the starting point for GIT. See [25, 44] for details.

Moduli space of binary forms

Example 3.5 A binary form of degree n is a polynomial in two variables defining n zeros counted with multiplicity. When rescaling a form the zeros remain the same, then to classify forms we will take the GIT quotient of the projective space of degree n polynomials in two variables up to the action of changing coordinates.

Let n be an integer and consider the set of homogeneous polynomials of degree n in two variables with coefficients in $\mathbb{C}$:

$$V_n = \left\{ f(X, Y) = a_0 Y^n + a_1 XY^{n-1} + a_2 X^2 Y^{n-2} + \cdots + a_{n-1} X^{n-1} Y + a_n X^n : a_i \in \mathbb{C} \right\}.$$

Let $\mathbb{P}(V_n)$ be its projectivization. The zeros of an element $\overline{f} \in \mathbb{P}(V_n)$ define n points in $\mathbb{P}^1_{\mathbb{C}}$ counted with multiplicity, up to the action of the group $G = \mathrm{SL}(2, \mathbb{C})$:

$$\begin{aligned}
\mathrm{SL}(2, \mathbb{C}) \times \mathbb{P}(V_n) &\longrightarrow \mathbb{P}(V_n) \\
(A, \overline{f}) &\longmapsto A\overline{f} := \overline{f}\left(A^{-1}(X, Y)^t\right)
\end{aligned}$$

The orbit space $\mathbb{P}(V_n)/G$ is not a variety because it is not Hausdorff. To see this, let $\overline{f}$ and $\overline{g}$ be given by

$$\overline{f}(X, Y) = X^n , \quad \overline{g}(X, Y) = X^n + X^{n-1}Y \in V_n ,$$

respectively. The orbits of these two elements are disjoint because the only root of $\overline{f}$ is $[0 : 1]$ counted with multiplicity n, and $\overline{g}$ has two roots: $[0 : 1]$ counted with multiplicity $n - 1$ and the simple root $[1 : -1]$. Let

$$h(t) = \begin{pmatrix} t & 0 \\ 0 & t^{-1} \end{pmatrix}$$

be a curve of elements in $\mathrm{SL}(2, \mathbb{C})$ and define

$$\overline{g}_t(X, Y) := h(t)\overline{g}(X, Y) = \overline{g}(h(t)^{-1}(X, Y)^t) =$$

$$\overline{g}(t^{-1}X, tY) = t^{-n}X^n + t^{-n+2}X^{n-1}Y \in \mathbb{P}(V_n) .$$

For each t, $\overline{g}_t$ defines the following element in $\mathbb{P}(V_n)$, by rescaling:

$$\overline{g_t}(X, Y) = t^{-n}X^n + t^{-n+2}X^{n-1}Y \sim X^n + t^2 X^{n-1}Y .$$

Then, note that when t goes to 0, $\overline{g_t}$ tends to $X^n = \overline{f}$; therefore, $\overline{f}$ lies in the closure of the orbit of $\overline{g}$ and the orbit space is not Hausdorff.

In order to construct a GIT quotient we are going to apply the Hilbert–Mumford criterion in Theorem 3.3. There exists a basis of $\mathbb{C}^2$ such that the 1-parameter subgroups of $\mathrm{SL}(2, \mathbb{C})$ can be diagonalized. Then, up to conjugation they are classified by integers $k > 0$ and represented by matrices like

$$\rho_k(t) = \begin{pmatrix} t^{-k} & 0 \\ 0 & t^k \end{pmatrix}$$

such that, if we write $\overline{f}(X, Y) = \displaystyle\sum_{i=0}^{n} a_i X^i Y^{n-i}$, the action of ρ_k is given by

$$\rho_k(t)\overline{f}(X, Y) = \overline{f}(\rho_k(t)^{-1}(X, Y)^t) = \overline{f}(t^k X, t^{-k} Y) = \sum_{i=0}^{n} a_i t^{k(2i-n)} X^i Y^{n-i} .$$

The limit is

$$\overline{f_0} := \lim_{t \to 0} \rho_k(t)\overline{f} = a_{i_0} X^{i_0} Y^{n-i_0} \, ,$$

where i_0 is the minimum index i such that $a_i \neq 0$. For example, if $\overline{f}(X, Y) = XY^4 + X^3Y^2 + X^5$, then

$$\rho_k(t)\overline{f}(X, Y) = t^{-3k} XY^4 + t^k X^3 Y^2 + t^{5k} X^5 \sim XY^4 + t^{4k} X^3 Y^2 + t^{8k} X^5 \, ,$$

which tends to $XY^4 = \overline{f_0}$ when t goes to 0. Note that the weight of the 1-parameter subgroup ρ_k is given by

$$\vartheta(\overline{f}, \rho_k) = k(2i_0 - n)$$

which, in the example, gives

$$\vartheta(\overline{f}, \rho_k) = -3k \, .$$

The Hilbert–Mumford criterion in Theorem 3.3 states that a point $\overline{f}$ is unstable if there exists a 1-parameter subgroup such that this weight is positive. Also recall that, up to conjugation in $\mathrm{SL}(2, \mathbb{C})$ (or change of homogeneous coordinates $[x : y]$), all 1-parameter subgroups are of the diagonal form ρ_k with $k > 0$. Therefore, $\overline{f}(X, Y) = \sum_{i=0}^{n} a_i X^i Y^{n-i}$ is unstable if, after a change of coordinates, there exists a 1-parameter subgroup ρ_k such that $k(2i_0 - n) > 0$ where i_0 is the minimum index with $a_{i_0} \neq 0$. Given that

$$k(2i_0 - n) > 0 \iff i_0 > \frac{n}{2} \, ,$$

it is equivalent to say that $\overline{f}$ is unstable if and only if $\overline{f}$ has a root of multiplicity greater that $n/2$.

In the example of the polynomial $\overline{f}(X, Y) = XY^4 + X^3Y^2 + X^5$, the weight is

$$\vartheta(\overline{f}, \rho_k) = -3k < 0 \, ,$$

and the lifted orbit

$$\rho_k(t)\overline{f}(X, Y) = t^{-3k} XY^4 + t^k X^3 Y^2 + t^{5k} X^5$$

tends to infinity when t goes to 0; hence, this 1-parameter subgroup does not contradict the stability of the point $\overline{f}$. Indeed, it will occur the same with all 1-parameter subgroups as it is easy to check, because $\overline{f}$ has no root of multiplicity ≥ 3, where $3 > \frac{n}{2}$ with $n = 5$. However, the point $\overline{g}(X, Y) = X^3Y^2$ will be acted by ρ_k as

$$\rho_k(t)\overline{g}(X, Y) = t^k X^3 Y^2 \ ,$$

which goes to 0 when t goes to 0. Hence, 0 is in the closure of the lifted orbit and the weight is $\vartheta(\overline{g}, \rho_k) = k > 0$, then the point is unstable. Indeed, $\overline{g}$ has a root with multiplicity 3 (in these coordinates the root is $[1 : 0]$).

Observe that, if n is odd, we cannot have $i_0 = \frac{n}{2}$, hence we cannot have strictly semistable points and all the semistable ones are stable.

If n is even, we can observe the S-equivalence phenomenon. Let $n = 4$ and consider the points

$$\overline{f}(X, Y) = X^2 Y^2 + X^3 Y + X^4 \quad \text{and} \quad \overline{g}(X, Y) = X^2 Y^2 \ .$$

By the same argument that we used to show that the orbit space is not Hausdorff, it is clear that $\overline{f}$ (with two roots equal and the other two different) and $\overline{g}$ (with two roots pairwise equal) do not lie in the same orbit, but $\overline{g}$ lies in the closure of the orbit of $\overline{f}$. Hence the two points are S-equivalent. To determine which one is the unique polystable orbit within this equivalence class, we can use any 1-parameter subgroup ρ_k with $k > 0$, which acts on the fiber of the limit point (common to $\overline{f}$ and $\overline{g}$ and, indeed, equal to $\overline{g}$) with weight zero, and conclude that $\overline{g}$ is the polystable orbit.

> The moduli space of configurations of n points in the projective line is the same that the moduli space of n-gons (c.f. [55]) if we consider the isomorphism $\mathbb{P}^1_{\mathbb{C}} \simeq S^2$ and see points in $\mathbb{P}^1_{\mathbb{C}}$ as unit length vectors. A configuration of points will be unstable if there exists a point with multiplicity more than half the number of points, equivalently a polygon will be unstable if there are any of the vectors repeated more than half the times. It can be shown that an unstable polygon does not close, in the sense that, after any change of coordinates by $SL(2, \mathbb{C})$, the sum of the vectors is not zero (see Fig. 3.3).

3.3 Symplectic Stability

In this section we will sketch the symplectic reduction procedure, giving another perspective of the stability picture. The Kempf–Ness theorem will be the link in between the two sides. Let us begin by reviewing the basics about symplectic geometry.

Let (X, ω) be a *symplectic manifold*, where X is a smooth manifold and $\omega \in \Omega^2(X)$ is a closed nondegenerate 2-form called a *symplectic form*. Two symplectic varieties (X_1, ω_1) and (X_2, ω_2) are *symplectomorphic* if there exists a diffeomorphism

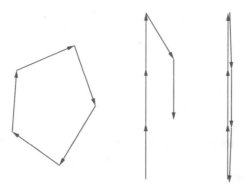

Fig. 3.3 Stability for binary forms in the fashion of polygons. Under the isomorphism $\mathbb{P}^1_\mathbb{C} \simeq S^2$, points in $\mathbb{P}^1_\mathbb{C}$ correspond to unit length vectors, then the zeros of a degree n binary form yield n unit vectors in S^2 which are the sides of a (not necessarily closed) n-gon. The first two figures represent, for $n = 5$, a stable polygon which closes and an unstable polygon which does not close, respectively. The third figure shows a strictly semistable polygon for $n = 6$, where there are two distinct vectors, repeated three times each

$$\alpha : X_1 \longrightarrow X_2 \quad \text{such that} \quad \alpha^* \omega_2 = \omega_1 \ .$$

By Darboux's theorem every symplectic manifold is locally symplectomorphic to $\mathbb{R}^{2n}$ equipped with the standard symplectic 2-form $\sum_{i=1}^{n} dq_i \wedge dp_i$.

Given a symplectic manifold (X, ω), let $\mathrm{Symp}(X, \omega) \subset \mathrm{Diff}(X)$ be the group of symplectomorphisms and let $\mathrm{Vect}^s(X) \subset \mathrm{Vect}(X)$ be the Lie subalgebra of symplectic vector fields $v \in \mathrm{Vect}(X)$ such that the Lie derivative of ω along v vanishes:

$$\mathcal{L}_v \omega = d\,(\iota_v \omega) = 0 \ ,$$

where ι_v is the interior product of differential forms, that is, the contraction of a differential form with a vector field, $\iota_v(\omega)(\cdot) = \omega(v, \cdot)$.

Consider the Lie group $S^1 = U(1)$ acting on a symplectic manifold (X, ω). We say that the action is *symplectic* if there exists a family of symplectomorphisms

$$\alpha_t \in \mathrm{Symp}(X)\,, t \in S^1, \ \text{such that } \alpha_{t+s} = \alpha_t \circ \alpha_s\,, \ \text{for every } t, s \in S^1 \ .$$

It can be seen that the associated vector fields

$$v_t = \frac{d}{dt}\alpha_t \circ \alpha_t^{-1} \in \mathrm{Vect}(X)$$

are all the same v, independently of t. The 1-form $\iota_v\omega$ is closed and then $v \in$ Vect$^s(X)$ is a symplectic vector field. When it is, moreover, exact, that is, $\iota_v\omega = dH$ with $H \in \mathcal{E}_X$, we say that it is a *Hamiltonian action* and the function

$$H : X \longrightarrow \mathbb{R}$$

is called *moment map*. This function is well defined up to a real constant, and it has the property that the level sets $H^{-1}(a)$, $a \in \mathbb{R}$, are S^1-invariant because $dH(v) = \omega(v, v) = 0$. And its critical points correspond to fixed points of the S^1-action because $x \in X$ is fixed if and only if $v_x = 0$, which is equivalent to $dH = \iota_v\omega = \omega(v, \cdot)$ being zero at $x \in X$.

Note that a smooth function $H \in \mathcal{E}_X$ defines a symplectic vector field v_H by the condition $\iota_{v_H}\omega = dH$, because $d(\iota_{v_H}\omega) = d(dH) = 0$. Observe that the image of $\mathcal{E}_X$ lies in the subalgebra Vect$^s(X)$ of symplectic vector fields. In local Darboux coordinates, v_H is given by

$$v_H = \sum_{i=1}^{n} \frac{\partial H}{\partial p_i}\frac{\partial}{\partial q_i} - \frac{\partial H}{\partial q_i}\frac{\partial}{\partial p_i} \ ,$$

from which we can see, by remembering the Hamilton equations, how symplectic geometry gives the natural framework for mechanics. We call $d\mathcal{E}_X = \text{Vect}^H(X)$ the *Hamiltonian* vector fields. Given that the kernel of d are the constant functions, the Lie algebra of the Hamiltonian automorphisms is $\mathcal{E}_X/\mathbb{R}$.

In general, let K be a compact connected Lie group acting on a symplectic manifold (X, ω) and let $\mathfrak{k}$ be its Lie algebra. We say that the action is *symplectic* if it preserves the symplectic form, that is, the diffeomorphisms are symplectomorphisms:

$$k_X : x \longmapsto kx \in \text{Symp}(X, \omega) \ , \ \forall k \in K \ .$$

We say that the action is *Hamiltonian* if the map $\mathfrak{k} \to \text{Vect}(X)$ (which sends an element $v \in \mathfrak{k}$ to the corresponding vector field in X) lifts, equivariantly by the action of K, to a Hamiltonian vector field v_H with $H \in \mathcal{E}_X$, such that $\iota_{v_H}\omega = dH$. In this case, we can define a *moment map*

$$\psi : X \longrightarrow \mathfrak{k}^\vee$$

by the condition

$$\iota_v\omega = -d\langle\psi, v\rangle \ , \ \forall v \in \mathfrak{k} \ .$$

Given that the Lie algebra of the Hamiltonian automorphisms is $\mathcal{E}_X/\mathbb{R}$, we can choose each element v_H up to a constant; hence, the lifting condition means that we choose these constants in such a way that ψ is K-equivariant (by the coadjoint action on the right-hand side). Therefore, given a Hamiltonian K-action, the moment map is unique up to the addition of a central element of $\mathfrak{k}^{\vee}$.

In the following, let $X \subset \mathbb{P}^n_{\mathbb{C}}$ be a projective variety with an action of a compact connected Lie algebraic group K, whose complexified group is $G = K_{\mathbb{C}}$ (which is, hence, reductive), such that

$$G \subset \mathrm{GL}(n+1, \mathbb{C}) \quad \text{and} \quad K \subset \mathrm{U}(n+1) .$$

Suppose that K acts on $\mathbb{P}^n_{\mathbb{C}}$ by preserving the almost-complex structure J and the Fubiny-Study metric g, K preserves the natural symplectic structure $\omega = g(\cdot, J\cdot)$. In this case there is a natural moment map which, for $K = \mathrm{U}(n)$ and identifying its Lie algebra $\mathfrak{u}(n)$ with its dual via the inner product $\langle A, B \rangle = \mathrm{tr}(A^*B)$, is given by

$$\psi : \mathbb{P}^n_{\mathbb{C}} \to \mathfrak{u}(n) , \quad \psi(z) = \frac{i}{2|z|^2} zz^* , \tag{3.1}$$

up to addition of a central element which in this case is a constant. If we consider the natural action of $\mathrm{U}(n)$ on $\mathbb{C}^n$, endowed with its canonical symplectic form, the moment map simplifies to

$$\psi : \mathbb{C}^n \to \mathfrak{u}(n) , \quad \psi(z) = \frac{i}{2} zz^* . \tag{3.2}$$

When we have a diagonal action on a product of symplectic varieties, it can be proved that the moment map is the sum of the respective moment maps.

The different moment maps for a given action correspond to the different polarizations and linearizations of the action from the Geometric Invariant Theory side. If the symplectic form ω is integral, meaning that its cohomology class lies in the subgroup

$$H^2(X, \mathbb{Z})/\,\text{torsion} \subset H^2(X, \mathbb{R}) ,$$

then $2\pi i\omega$ is the curvature of a line bundle L with an hermitian structure and a unitary connection, and the isometries of L preserving the connection cover the Hamiltonian automorphisms on X (see Sects. 4.3 and 4.4).

In the projective case, this cohomology class is always integral; hence, we can develop this *prequantization* to restrict to a discrete number of different moment maps, associated to the GIT linearizations.

In the symplectic setting we state the following notion of stability.

Definition 3.3 Let (X, ω) be a projective variety with the symplectic form ω coming from the Fubini-Study metric, endowed with a Hamiltonian K-action. Let

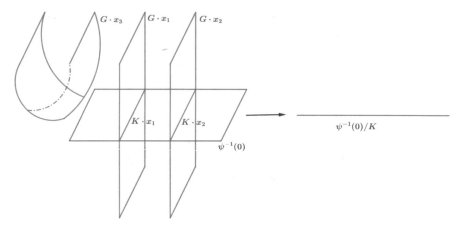

Fig. 3.4 G-orbits and K-orbits of a symplectic action and preimage of 0 under the moment map ψ. The points x_1 and x_2 are ψ-stable while x_3 is ψ-unstable

ψ be a moment map for this action. Let $x \in X$ and let us denote by $G \cdot x$ its orbit by the complexified group $G = K_{\mathbb{C}}$.

- x is ψ-*semistable* if $\overline{G \cdot x} \cap \psi^{-1}(0) \neq \emptyset$.
- x is ψ-*polystable* if $G \cdot x \cap \psi^{-1}(0) \neq \emptyset$.
- x is ψ-*stable* if x is ψ-polystable and the stabilizer of x under G is finite.
- x is ψ-*unstable* if $\overline{G \cdot x} \cap \psi^{-1}(0) = \emptyset$.

The notions of GIT stability and ψ-stability will be equivalent to the Kempf–Ness theorem (Fig. 3.4).

Theorem 3.4 (Kempf–Ness Theorem [57]) *Let (X, ω) be a projective variety with the symplectic form ω coming from the Fubini-Study metric, endowed with a Hamiltonian K-action. Let ψ be a moment map for this action. A G-orbit is GIT polystable if and only if it contains a zero of the moment map ψ. A G-orbit is GIT semistable if and only if its closure contains a zero of the moment map ψ, and this zero lies in the unique GIT polystable orbit in the closure of the original orbit.*

We will make some considerations to sketch the proof of the Kempf–Ness theorem.

Let $(X, L = O_X(1))$ be a projective polarized variety such that $X \hookrightarrow \mathbb{P}(H^0(X, L)^{\vee})$, and choose an hermitian metric on L inducing a connection (see Sect. 4.3.2) with curvature $2\pi i\omega$. Lift a point $x \in X$ to $\hat{x}$ in the fiber L_x^{-1} and consider the functional norm $\|\hat{x}\|$. A metric in $H^0(X, L)^{\vee}$ induces a metric in the total space of L^{-1} where $\|\hat{x}\|$ is the norm in the vector space where the affine cone $\hat{X}$ lives.

For each $\hat{x}$, define the Kempf–Ness function:

$$\Psi_{\hat{x}} : \mathfrak{k} \longrightarrow \mathbb{R}, \quad v \longmapsto \frac{\log \|\exp(iv)\hat{x}\|^2}{2}.$$

The 1-parameter subgroups encoding GIT stability by the Hilbert–Mumford criterion can be thought as different directions in the G-orbit, hence different elements of the Lie algebra of G, which decomposes as $\mathfrak{g} = \mathfrak{k} \oplus i\mathfrak{k}$. To study how this function varies along 1-parameter subgroups we calculate

$$\partial_\lambda \Psi_{\hat{x}}(v) = \frac{d}{dt}\bigg|_{t=0} \frac{\log \|\exp(i(v + t\lambda))\hat{x}\|^2}{2} =$$

$$\frac{\langle i\lambda \exp(iv)\hat{x}, \exp(iv)\hat{x}\rangle}{\langle \exp(iv)\hat{x}, \exp(iv)\hat{x}\rangle} = 2\psi\,((\exp iv)x)\,(\lambda),$$

which can be expressed by saying that the Kempf–Ness function is an integral of the moment map. If we calculate the second derivative, we obtain

$$\partial_\eta \partial_\lambda \psi_{\hat{x}}(v) = 2\langle \mathcal{L}_{J\eta}\psi\,((\exp iv)x), \lambda\rangle =$$

$$(\omega(\lambda, J\eta))\,(\exp(iv)x) = g(\lambda, \eta)\,(\exp(iv)x),$$

which is non-negative, since g is a Riemannian metric.

Hence, the Kempf–Ness function is convex, attaining a minimum at the zeros of the function $\psi\,((\exp iv)x)$ which are the zeros of the moment map. This way, x is ψ-polystable if and only if $\Psi_{\hat{x}}$ attains a minimum. If the Kempf–Ness function is bounded from below, it does not necessarily attain a minimum, but, if it does asymptotically, it means that the closure of the G-orbit of the point contains a zero of the moment map and the point is ψ-semistable.

The unstable points will be those for which the Kempf–Ness function is not bounded from below or, equivalently, the orbit under the complexified group does not intersect the zeros of the moment map. The GIT unstable points are those $x \in X$ for which $0 \in \overline{G \cdot \hat{x}}$, where $\hat{x}$ lies over x in the affine cone. From the definition of the Kempf–Ness function $\Psi_{\hat{x}}$ in terms of the logarithm, $0 \in \overline{G \cdot \hat{x}}$ will be equivalent to $\Psi_{\hat{x}}$ not being bounded by below, which is equivalent to the ψ-unstability of x.

From this, the *symplectic quotient* construction is due to Meyer and Marsden-Weinstein:

Theorem 3.5 ([64, 68]) *Let (X, ω) be a symplectic manifold endowed with a Hamiltonian action of a compact connected Lie group K. If ψ is a moment map for this action, and K acts freely and properly on $\psi^{-1}(0)$, the quotient $\left(\psi^{-1}(0)/K, \omega_0\right)$ is a smooth symplectic manifold with $i^*\omega = p^*\omega_0$, where $i : \psi^{-1}(0) \to X$ is the inclusion and $p : \psi^{-1}(0) \to \psi^{-1}(0)/K$ the projection, respectively.*

By the Kempf–Ness theorem, we will have the following bijection relating the GIT and the symplectic quotients, which is indeed a homeomorphism:

$$\psi^{-1}(0)/K \simeq X^{ps}/G = X^{ss} /\!/ G.$$

3.4   Examples

Next, we will calculate the moment map for the examples studied from the algebraic setting and check that the Kempf–Ness theorem holds in these cases.

Projective space as a symplectic quotient

Example 3.6 Recall the action of Example 3.2 which is a circle action with

$$K = S^1 = U(1) \subset G = \mathbb{C}^* = \mathrm{GL}(1, \mathbb{C}) \ .$$

Using (3.2) and the fact that a diagonal circle action yields, as a moment map, the sum of the respective moment maps, we get

$$\psi : \quad \mathbb{C}^{n+1} \quad \longrightarrow \quad \mathfrak{k}^\vee = \mathfrak{u}(1)^\vee \simeq \mathbb{R}$$
$$(z_0, \dots, z_n) \longmapsto \tfrac{i}{2}(|z_0|^2 + \cdots |z_n|^2 - a)$$

where a comes from a central element of $\mathfrak{k}^\vee \simeq \mathbb{R}$, that is, $a \in \mathbb{R}$. If we add the condition that the lifted action of $\mathfrak{u}(1)$ descends to an action of the group $K = U(1)$ on the trivial line bundle, we have that a is indeed an integer. The different $a \in \mathbb{Z}$ correspond to the integers p of the different characters in Example 3.2.

- If $a > 0$, the origin is ψ-unstable because its orbit does not intersect $\psi^{-1}(0)$. All the other orbits intersect $\psi^{-1}(0)$ at some $(z_0, \dots, z_n)$ such that $\sum_{i=0}^{n} |z_i|^2 = a$; hence, all rays are ψ-polystable (indeed ψ-stable) and the quotient is the expected projective space $\mathbb{P}^n_{\mathbb{C}}$.
- If $a < 0$, there are no $\mathbb{C}^*$-orbits intersecting $\psi^{-1}(0)$, not even in the closure; hence, all points are ψ-unstable as well as they are GIT unstable.
- If $a = 0$, the origin in $\mathbb{C}^{n+1}$ is ψ-polystable because its orbit intersects $\psi^{-1}(0)$. All the other orbits are ψ-semistable but not ψ-polystable because their closures intersect $\psi^{-1}(0)$. The origin is in the closure of all orbits; hence, it is the unique polystable point in the equivalence class and the symplectic quotient is a singleton.

Binary forms as a symplectic quotient

Example 3.7 Now we recall the classification of configurations of n points in $\mathbb{P}^1_{\mathbb{C}}$, from Example 3.5. Identify each $\overline{f} \in \mathbb{P}(V_n)$ with its set of n zeros counted with multiplicity and, by the homeomorphism $\mathbb{P}^1_{\mathbb{C}} \simeq S^2$, identify them with n vectors in the unit sphere. The compact group now is $SO(3, \mathbb{R}) \subset SL(2, \mathbb{C})$, acting diagonally on $(S^2)^n$ by rotations. The Lie algebra of $SO(3, \mathbb{R})$ is $\mathfrak{so}(3, \mathbb{R}) \simeq \mathbb{R}^3$, and the moment map in this case is just the sum of the inclusions of each vector in $\mathbb{R}^3$, hence given by (see [93]):

$$\psi : \quad (S^2)^n \quad \longrightarrow \quad \mathfrak{so}(3, \mathbb{R})^\vee \simeq \mathbb{R}^3$$
$$(v_1, \ldots, v_n) \longmapsto v_1 + \cdots + v_n .$$

Then, a configuration of points will be ψ-semistable if and only if the associated n-tuple of vectors $(v_1, \ldots, v_n)$ (up to the action of the complexified group $SL(2, \mathbb{C})$) verifies $\sum_{i=1}^{n} v_i = 0$, which is the equivalent to say that a "polygon closes," identifying this problem with the moduli space of polygons (see [55] and Fig. 3.3).

Since the Kempf–Ness theorem asserts that ψ-stability is equivalent to GIT stability, this means the following. A configuration of n points in $\mathbb{P}^1_{\mathbb{C}}$ can be moved by an element of $SL(2, \mathbb{C})$ such that the corresponding n-tuple of vectors in S^2 (counted with multiplicity) have its center of mass at the origin, if and only if there is no point with multiplicity greater than half the total, which means that the point is semistable.

In case n is even, we can have a point with multiplicity exactly half the total and recall that this means the point is GIT semistable but not stable. The polynomials

$$\overline{f} = X^2 Y^2 + XY^3 + Y^4 \quad \text{and} \quad \overline{g} = X^2 Y^2$$

verify that $\overline{g} \in \overline{SL(2, \mathbb{C}) \cdot \overline{f}}$. The polynomial $\overline{g}$ defines a configuration with only two points, each of them with the same multiplicity equaling to half the total. For example, for these polynomials the points are $[1 : 0]$ and $[0 : 1]$, which in S^2 can be thought as the opposite vectors $(0, 0, 1)$ and $(0, 0, -1)$. Then $\overline{g}$ is the only polystable orbit in the closure of the orbit of $\overline{f}$ which defines a (degenerate) configuration of vectors in the unit sphere with its center of mass at the origin, therefore

$$\psi^{-1}(0) \cap G \cdot \overline{g} \neq \emptyset \quad \text{and} \quad \psi^{-1}(0) \cap \overline{G \cdot \overline{f}} \neq \emptyset$$

but

$$\psi^{-1}(0) \cap G \cdot \overline{f} = \emptyset ,$$

meaning that $\overline{f}$ is ψ-semistable but not ψ-polystable, and $\overline{g}$ is ψ-polystable. By visualizing polygons, this situation in general corresponds to the degenerate polygon with $n/2$ vectors equal to v and the other $n/2$ equal to $-v$ lying on a line, which can only appear for n even (see the third picture in Fig. 3.3). This limit point corresponds to the polystable orbit with stabilizer $\mathbb{C}^*$.

Grassmannian as GIT and symplectic quotients

Example 3.8 Let us construct the Grassmannian $\mathrm{Gr}(r, n)$ as a GIT quotient and as a symplectic quotient. The points of the Grassmannian of r-dimensional vector subspaces of $\mathbb{C}^n$ correspond to injective homomorphisms from $\mathbb{C}^r$ to $\mathbb{C}^n$, up to a change of basis. This change of basis is encoded by considering the projective space $\mathbb{P}\left(\mathrm{Hom}(\mathbb{C}^r, \mathbb{C}^n)\right)$, where two linear maps differing by scalar multiplication define the same vector subspace, and by the action of $\mathrm{SL}(r, \mathbb{C})$ which are the changes of frame with determinant one.

We will obtain the Grassmannian as the GIT quotient of $\mathbb{P}\left(\mathrm{Hom}(\mathbb{C}^r, \mathbb{C}^n)\right)$ by the action of $\mathrm{SL}(r, \mathbb{C})$. Let

$$\mathrm{SL}(r, \mathbb{C}) \times \mathbb{P}\left(\mathrm{Hom}(\mathbb{C}^r, \mathbb{C}^n)\right) \longrightarrow \mathbb{P}\left(\mathrm{Hom}(\mathbb{C}^r, \mathbb{C}^n)\right)$$

be the group action such that

$$[A] \cdot g^{-1} \quad \text{for} \quad [A] \in \mathbb{P}\left(\mathrm{Hom}(\mathbb{C}^r, \mathbb{C}^n)\right) \,, \quad g \in \mathrm{SL}(r, \mathbb{C}) \,.$$

The affine cone of $\mathbb{P}\left(\mathrm{Hom}(\mathbb{C}^r, \mathbb{C}^n)\right)$ can be identified with $\mathrm{Hom}(\mathbb{C}^r, \mathbb{C}^n)$ and we lift the action $[A] \cdot g^{-1}$ to $A \cdot g^{-1}$ where $A \in \mathrm{Hom}(\mathbb{C}^r, \mathbb{C}^n)$ lies over $[A]$ in the affine cone.

Let us prove that $[A] \in \mathbb{P}\left(\mathrm{Hom}(\mathbb{C}^r, \mathbb{C}^n)\right)$ is GIT stable if and only if $A \in \mathrm{Hom}(\mathbb{C}^r, \mathbb{C}^n)$ has rank r. Therefore the $\mathbb{P}\left(\mathrm{Hom}(\mathbb{C}^r, \mathbb{C}^n)\right)^s$ are the injective homomorphisms and the GIT quotient is $\mathrm{Gr}(r, n) = \mathbb{P}\left(\mathrm{Hom}(\mathbb{C}^r, \mathbb{C}^n)\right)^s / \mathrm{SL}(r, \mathbb{C})$.

If $\mathrm{rk}(A) < r$, pick a basis $\{v_1, \ldots, v_r\}$ of $\mathbb{C}^r$ such that $v_1 \in \mathrm{Ker}\, A$. Choose a 1-parameter subgroup $\rho : \mathbb{C}^* \to \mathrm{SL}(r, \mathbb{C})$ adapted to the basis such that it has the diagonal form:

$$\rho(t) = \begin{pmatrix} t^{r-1} & 0 & \cdots & 0 \\ 0 & t^{-1} & \cdots & 0 \\ \vdots & & \ddots & \vdots \\ 0 & 0 & \cdots & t^{-1} \end{pmatrix}$$

Then:

$$A \cdot \rho^{-1}(t) = \begin{pmatrix} 0 & * & \cdots & * \\ 0 & * & \cdots & * \\ \vdots & \vdots & & \vdots \\ 0 & * & \cdots & * \end{pmatrix} \cdot \begin{pmatrix} t^{1-r} & 0 & \cdots & 0 \\ 0 & t^1 & \cdots & 0 \\ \vdots & & \ddots & \vdots \\ 0 & 0 & \cdots & t^1 \end{pmatrix} = \begin{pmatrix} 0 & t \cdot * & \cdots & t \cdot * \\ 0 & t \cdot * & \cdots & t \cdot * \\ \vdots & \vdots & & \vdots \\ 0 & t \cdot * & \cdots & t \cdot * \end{pmatrix} = t \cdot A \xrightarrow{t \to 0} 0 \,,$$

hence ρ acts on A as multiplying by t, this is with weight $\vartheta\left([A], \rho\right) = 1 > 0$ and therefore $[A]$ is GIT unstable.

Conversely, let A be of rank r and let ρ be a nontrivial 1-parameter subgroup of $SL(r, \mathbb{C})$. There exists a basis of $\mathbb{C}^r$ and a splitting $\mathbb{C}^n \simeq \mathbb{C}^r \oplus \mathbb{C}^{n-r}$ such that A is the inclusion of the first factor in this splitting and ρ is of the diagonal form:

$$\rho(t) = \begin{pmatrix} t^{\rho_1} & 0 & 0 & \cdots & 0 \\ 0 & t^{\rho_2} & 0 & \cdots & 0 \\ 0 & 0 & t^{\rho_3} & \cdots & 0 \\ \vdots & \vdots & & \ddots & \vdots \\ 0 & 0 & 0 & \cdots & t^{\rho_r} \end{pmatrix}$$

with $\sum_{i=1}^{r} \rho_i = 0$ and assume $\rho_1 \geq \rho_2 \geq \cdots \geq \rho_r$, then $\rho_1 > 0$. The action $A \cdot \rho(t)^{-1}$ multiplies the ith column of the matrix A by $t^{-\rho_i}$ and the first column of A has, at least, one non-zero entry, then the minimum weight is $\vartheta([A], \rho) = -\rho_1 < 0$ and $[A]$ is GIT stable.

From the symplectic point of view, we have the action of the compact unitary group $U(r) \subset GL(r, \mathbb{C})$ on $\mathrm{Hom}(\mathbb{C}^r, \mathbb{C}^n)$ the same way, given by $A \cdot g^{-1}$. Considering the inner product $\langle A, B \rangle = \mathrm{tr}(A^*B)$ which identifies $\mathfrak{u}(n)^\vee$ with $\mathfrak{u}(n)$, a moment map for the action can be shown to be:

$$\psi : \mathrm{Hom}(\mathbb{C}^r, \mathbb{C}^n) \longrightarrow \mathfrak{u}(r)^\vee$$
$$A \longmapsto i(A^*A - I_r)$$

Hence, $\psi^{-1}(0)$ are those matrices such that, up to the action of $GL(r, \mathbb{C})$, verify $A^*A = I_r$, that is, matrices of linear maps congruent by $GL(r, \mathbb{C})$ to isometric embeddings, which happens if and only if the linear map is injective.

In general, we could have added a central element (in this case a scalar matrix) to the moment map to get $\psi(A) = i(A^*A - a \cdot I_r)$. If $a > 0$, we obtain the same result. If $a = 0$, the quotient is a single point, and if $a < 0$, all points are ψ-unstable. This can be put in correspondence with different linearizations in the GIT problem, as in Example 3.2.

3.5 Maximal Unstability

After studying the relationship between GIT stability and symplectic stability by the Kempf–Ness theorem, in this section we will focus on the unstable locus. We will classify the unstable points by degrees of unstability and will check that this notion agrees when considered from both points of view.

The moment map $\psi : X \to \mathfrak{k}^{\vee}$ is invariant by the adjoint action of the compact group K but not by the action of its complexified group $G = K_{\mathbb{C}}$. If we choose an inner product $\langle \cdot , \cdot \rangle$ in $\mathfrak{k}$, invariant by K, we can identify $\mathfrak{k}^{\vee}$ with $\mathfrak{k}$ and define the function

$$\|\psi\| : X \longrightarrow \mathbb{R} , \quad x \longmapsto \|\psi(x)\| := \langle \psi(x), \psi(x) \rangle ,$$

to which we will refer as the *moment map square*. Recall that the Kempf–Ness function is an integral of the moment map. The ψ-unstable points are those x for which $\psi(g \cdot x)$ does not achieve zero as a limit point, for $g \in G$, hence the Kempf–Ness function for these points is unbounded.

Given $x \in X$, define the function

$$\Xi_x(g) := \|\psi(g \cdot x)\| , \quad g \in G .$$

The function Ξ_x is a Morse-Bott function and it takes some infimum value $m_x \geq 0$ at the critical set. The idea is that there exists a direction of maximal descent for the negative gradient flow of the Kempf–Ness function, directions thought as cosets in G/K, minimizing the moment map square, that is, the function Ξ_x (c.f. [59] and [23]). Then, the G-orbit of a ψ-unstable point x does not achieve the preimage $\Xi_x^{-1}(0)$, but it does achieve, in their closure, the set $\Xi_x^{-1}(m_x)$ for some positive number m_x (see moment limit theorem [23, Theorem 6.4] and generalized Kempf existence theorem [23, Theorem 11.1]). Of course, for the ψ-semistable ones, this infimum m_x is zero.

From the algebraic point of view, recall that a point x is GIT unstable if there exists a 1-parameter subgroup ρ such that the weight $\vartheta(x, \rho)$ is positive. Recall that the number $\vartheta(x, \rho)$ is the weight that ρ is acting with on the fiber of the fixed limit point of $\rho(t)$ when t goes to zero. Having chosen the inner product $\langle \cdot , \cdot \rangle$ in $\mathfrak{k}$, it extends uniquely to an inner product in G. Considering the 1-parameter subgroups as directions given by elements in the Lie algebra $\mathfrak{g}$, it makes sense to define the *length* $\|\rho\|$ of a 1-parameter subgroup and, hence, the function

$$\Phi_x(\rho) := \frac{\vartheta(x, \rho)}{\|\rho\|} .$$

If x is GIT unstable, there exists a ρ such that $\Phi_x(\rho) > 0$. The result by Kempf [56] asserts that the supremum of the function Φ_x is attained at some unique ρ (up to conjugation by the parabolic subgroup of G defined by ρ), hence there exists a unique 1-parameter subgroup maximizing the Hilbert–Mumford criterion, or giving the maximal way to destabilize a GIT unstable point. The norm in the denominator serves to calibrate this maximal degree of unstability when rescaling, that is, multiplying the exponents of the 1-parameter subgroups by a scalar. We will revisit this idea in Chap. 5.

The principal result in [23] (c.f. [23, Theorem 13.1]) shows that, for an unstable point x, we have

$$\sup_{\rho \in \mathfrak{g}} \Phi_x(\rho) = \sup_{\rho \in \mathfrak{g}} \frac{\vartheta(x, \rho)}{\|\rho\|} = m_x = \inf_{g \in G} \Xi_x = \inf_{g \in G} \|\psi(g \cdot x)\| .$$

This is, the weight of the 1-parameter subgroup which maximally destabilizes a GIT unstable point x (after normalization), is the infimum of the moment map square over the G-orbit of a ψ-unstable point.

Maximal unstability for binary forms

Example 3.9 Let us go back to Example 3.5, the configurations of points in $\mathbb{P}^1_{\mathbb{C}}$. The group $\mathrm{SL}(2, \mathbb{C})$ is simple, then there is only one adjoint-invariant inner product up to multiplying by a scalar: the Killing norm (see Sect. 2.1.2). Following [56, p.305], we choose this natural inner product $\langle \cdot, \cdot \rangle$ such that the associated norm of $\rho_k(t) = \begin{pmatrix} t^{-k} & 0 \\ 0 & t^k \end{pmatrix}$ is equal to

$$\|\rho_k\| = \sqrt{\langle \rho_k, \rho_k \rangle} = \sqrt{\frac{1}{2} \mathrm{tr} \left(\mathrm{ad} \left((\rho_k)_* \frac{d}{dt} \right)^2 \right)} = \sqrt{\frac{1}{2} \mathrm{tr} \left(\begin{pmatrix} -k & 0 \\ 0 & k \end{pmatrix}^2 \right)} = k .$$

We did compute in Example 3.5 the weight of a 1-parameter subgroup ρ_k with exponents $-k$ and k in its diagonal form, obtaining $\vartheta\left(\overline{f}, \rho_k\right) = k(2i_0 - n)$. Here, recall that i_0 is the maximum number of points in $\mathbb{P}^1_{\mathbb{C}}$ which are equal. Then we get

$$\sup_{\rho \in \mathfrak{g}} \Phi_{\overline{f}}(\rho) = \sup_{\rho \in \mathfrak{g}} \frac{\vartheta\left(\overline{f}, \rho\right)}{\|\rho_k\|} = \frac{k(2i_0 - n)}{k} = 2i_0 - n ,$$

which is a positive number if $\overline{f}$ is GIT unstable.

Now, from the symplectic point of view, recall that we associate to each point in $\mathbb{P}^1_{\mathbb{C}}$ a vector in S^2 and the moment map is given by

$$\psi\left(\overline{f}\right) = v_1 + \cdots + v_n \in \mathbb{R}^3 ,$$

after identifying $\mathfrak{so}(3, \mathbb{R})^\vee \simeq \mathbb{R}^3$. The norm chosen in $\mathfrak{so}(3, \mathbb{R})$ can be identified with the usual Euclidean norm in $\mathbb{R}^3$.

Suppose that x is an unstable configuration, it defines $i_0 > \frac{n}{2}$ identical vectors in S^2. By changing the coordinates in $\mathbb{P}^1_{\mathbb{C}}$, we can consider that the configuration is given by a binary form

$$\overline{f}(X, Y) = a_{n-i_0} X^{n-i_0} Y^{i_0} + a_{n-i_0+1} X^{n-i_0-1} Y^{i_0+1} + \cdots + a_{n-1} XY^{n-1} + a_n Y^n ,$$

which can be moved in its SL(2, $\mathbb{C}$)-orbit by elements $g_t = \begin{pmatrix} t & 0 \\ 0 & t^{-1} \end{pmatrix}$ to obtain

$$g_t \overline{f} = \overline{f}\left(g_t^{-1}(X, Y)^t\right) = \overline{f}\left(t^{-1}X, tY\right) =$$

$$t^{2i_0-n} a_{n-i_0} X^{n-i_0} Y^{i_0} + t^{2i_0-n+2} a_{n-i_0+1} X^{n-i_0-1} Y^{i_0+1} + \cdots + t^{n-2} a_{n-1} X Y^{n-1} + t^n a_n Y^n \ .$$

We can multiply this by t^{n-2i_0} and still define the same form in the projective space:

$$a_{n-i_0} X^{n-i_0} Y^{i_0} + t^2 a_{n-i_0+1} X^{n-i_0-1} Y^{i_0+1} + \cdots + t^{2n-2i_0-2} a_{n-1} X Y^{n-1} + t^{2n-2i_0} a_n Y^n \ ,$$

which tends to $\overline{f_0} = a_{n-i_0} X^{n-i_0} Y^{i_0}$ when t goes to 0. The zeros of $\overline{f_0}$ are $[1 : 0]$ with multiplicity i_0 and $[0 : 1]$ with multiplicity $n-i_0$. When considering an isomorphism $\mathbb{P}^1_{\mathbb{C}} \simeq S^2$, we can associate these zeros to the vectors $(0, 0, 1)$ and $(0, 0, -1)$ in S^2. Hence, the calculation of the infimum of the moment map square is

$$\inf_{g \in G} \Xi_{\overline{f}} = \inf_{g \in G} \left\| \psi\left(g\overline{f}\right) \right\| \leq \inf_t \left\| \psi\left(g_t \overline{f}\right) \right\| =$$

$$\left| \sum_{i_0} (0, 0, 1) + \sum_{n-i_0} (0, 0, -1) \right| = \left| \sum_{2i_0-n} (0, 0, 1) \right| = 2i_0 - n =: m_{\overline{f}} \ .$$

And it is clear that the value $m_{\overline{f}}$ obtained is indeed the infimum, because the best we can do in order to get this infimum, once we have i_0 identical vectors in S^2, is to dispose the rest (up to action of SL(2, $\mathbb{C}$)) in the opposite direction, which we did by the curve of elements $g_t \in$ SL(2, $\mathbb{C}$).

As we observe,

$$\sup_{\rho \in \mathfrak{g}} \Phi_{\overline{f}}(\rho) = 2i_0 - n = \inf_{g \in G} \Xi_{\overline{f}} \ ,$$

therefore there are different levels of unstability, indexed by the numbers $m_{\overline{f}} = 2i_0 - n$. These correspond to binary forms with different number of identical roots, or to vectors in S^2 which do not close to form a polygon because they have different numbers of identical vectors, in all cases more than half of them.

Maximal unstability for Grassmannians

Example 3.10 Now we recall Example 3.8. Let $A \in \mathrm{Hom}(\mathbb{C}^r, \mathbb{C}^n)$ of rank $m < r$ lying over $[A] \in \mathbb{P}\left(\mathrm{Hom}(\mathbb{C}^r, \mathbb{C}^n)\right)$ in the affine cone. Hence $[A]$ is GIT unstable and there exists a basis $\{v_1, \ldots, v_r\}$ of $\mathbb{C}^r$ such that the vectors $v_1, \ldots, v_{r-m}$ span

Ker A. Adapted to this basis, the different diagonalized 1-parameter subgroups $\rho(t) \in SL(r, \mathbb{C})$ are given by

$$\rho(t) = \begin{pmatrix} t^{\rho_1} & 0 & \cdots & 0 \\ 0 & t^{\rho_2} & \cdots & 0 \\ \vdots & & \ddots & \vdots \\ 0 & 0 & \cdots & t^{\rho_r} \end{pmatrix}$$

with $\rho_1 \geq \rho_2 \geq \cdots \geq \rho_r$ and $\displaystyle\sum_{i=1}^{r} \rho_i = 0$. Then, $A \cdot \rho^{-1}(t) =$

$$\begin{pmatrix} \vdots & \vdots & \vdots & & \vdots \\ 0 \cdots 0 & * & \cdots & * \\ \vdots & \vdots & \vdots & & \vdots \end{pmatrix} \cdot \begin{pmatrix} t^{-\rho_1} & 0 & \cdots & 0 \\ 0 & t^{-\rho_2} & \cdots & 0 \\ \vdots & & \ddots & \vdots \\ 0 & 0 & \cdots & t^{-\rho_r} \end{pmatrix} = \begin{pmatrix} \vdots & \vdots & \vdots & & \vdots \\ 0 \cdots 0 & t^{-\rho_{r-m+1}} \cdot * & \cdots & t^{-\rho_r} \cdot * \\ \vdots & \vdots & \vdots & & \vdots \end{pmatrix}.$$

Hence, we observe that the weight of the Hilbert–Mumford criterion, that is, the minimal exponent multiplying a non-zero coordinate, is $\vartheta\left([A], \rho\right) = -\rho_{r-m+1}$. Therefore, in order to maximize this weight, keeping the condition that $\rho(t) \in SL(r, \mathbb{C})$ (hence all exponents sum up to 0), the maximal 1-parameter subgroups will be of the form:

$$\rho(t) = \begin{pmatrix} t^m & 0 & & \cdots & 0 & 0 \\ 0 & \ddots & & & 0 & 0 \\ & & t^m & & & \\ \vdots & & & t^{m-r} & & \vdots \\ 0 & 0 & & & \ddots & 0 \\ 0 & 0 & & \cdots & 0 & t^{m-r} \end{pmatrix} = \left(\begin{array}{c|c} t^m\, I_{r-m} & 0 \\ \hline 0 & t^{m-r}\, I_m \end{array} \right),$$

where t^m is repeated $r - m$ times and t^{m-r} is repeated m times. Then it is clear that for these 1-parameter subgroups we have $\vartheta\left([A], \rho\right) = r - m$. Note that we could have achieved the same maximal result by multiplying the exponents m and $m - r$ by the same positive constant hence, up to rescaling, the maximal weight will remain $r - m$. In other words,

$$\sup_{\rho \in \mathfrak{g}} \Phi_{[A]}(\rho) = \sup_{\rho \in \mathfrak{g}} \frac{\vartheta\left([A], \rho\right)}{\|\rho\|} = r - m .$$

From the symplectic side, recall that the moment map is given by

$$\psi(A) = i\left(A^* A - I_r\right) .$$

Having chosen the invariant product in $\mathfrak{u}(n)$ given by $\mathrm{tr}\,(A^*B)$, the moment map square is given by

$$\|\psi(A)\| = \mathrm{tr}\left(\left(A^*A - I_r\right)^* \left(A^*A - I_r\right)\right) = \mathrm{tr}\left(\left(A^*A - I_r\right)^2\right)$$

up to a constant. This constant is elated with the rescaling of the norm discussed before from the GIT point of view. By an element of $\mathrm{SL}(r, \mathbb{C})$ (or by change of basis) we can suppose that A^*A is a matrix with a diagonal block which is the identity I_m (of size the rank of A) and zeros elsewhere. Therefore, it is clear that

$$\inf_{g \in G} \Xi_{[A]} = \inf_{g \in G} \|\psi(g \cdot A)\| = \mathrm{tr}\left(\left(\begin{array}{c|c} 0 & 0 \\ \hline 0 & I_m \end{array}\right) - I_r\right)^2 = \mathrm{tr}\left(\begin{array}{c|c} I_{r-m} & 0 \\ \hline 0 & 0 \end{array}\right) = r - m \,,$$

which is equal to the quantity $\sup_{\rho \in \mathfrak{g}} \Phi_{[A]}(\rho)$. Hence, the different unstability levels are indexed by the complementary of the rank of A, being $m = r$ the case where the supremum and the infimum, respectively, achieve zero, as it has to be in the stable case.

Chapter 4
Moduli Space of Vector Bundles

The problem of classifying vector bundles is a very central story in geometry since the 1960s, with strong connections with other areas of mathematics and physics. The statement is to find a geometric structure with suitable properties (such as an algebraic variety), where each point corresponds to a holomorphic structure on a given smooth complex vector bundle.

When trying to perform this, the objects to classify happen to have different groups of automorphisms. This turns out to be a main issue because the moduli space attempts to collect all structures identifying the ones which are mathematically identical, this is, modding out by their automorphisms. Therefore, this prevents us from solving the problem of finding a moduli space for all vector bundles.

However, following an idea of Grothendieck, it is often the case that one can add a piece of data to the objects to classify, such that the only automorphism of these objects endowed with the additional data is the identity. The piece of data we add is encoded as the action of a group in a certain space. Geometric invariant theory, which was developed in [66] for this particular purpose as its main application, gives the solution to the quotient by the action of that group on the space, removing the additional data and yielding a moduli space as a GIT quotient.

The notion of stability that we define for vector bundles will distinguish between stable bundles, those with the smallest possible automorphism group and for which the solution of the moduli problem is the best possible (a *fine* moduli space parametrizing isomorphism classes), semistable bundles having a *coarse* moduli space (parametrizing S-equivalence classes), and unstable bundles left out of the classical moduli problem and that need to be dealt with by means of the Harder-Narasimhan filtration.

© The Author(s), under exclusive license to Springer Nature Switzerland AG 2021   59
A. Zamora Saiz, R. A. Zúñiga-Rojas, *Geometric Invariant Theory, Holomorphic Vector Bundles and the Harder-Narasimhan Filtration*, SpringerBriefs in Mathematics, https://doi.org/10.1007/978-3-030-67829-6_4

4.1 GIT Construction of the Moduli Space

Let us sketch the construction of a moduli space of semistable holomorphic vector bundles over a compact Riemann surface following geometric invariant theory. The reader can consult [25, 53, 62, 66, 71] for further details.

Let X be a smooth algebraic projective curve (or a compact Riemann surface) of genus g. Assume that X is embedded in a projective space by means of a very ample line bundle $O_X(1)$. Let E be a holomorphic vector bundle of rank $\mathrm{rk}(E) = r$, degree $\deg(E) = d$, and fixed determinant line bundle $\det(E) \simeq L$. Define the *slope* of a vector bundle by $\mu(E) := \frac{d}{r}$.

Definition 4.1 A vector bundle E is said to be *semistable* if for every nonzero proper subbundle $E' \subsetneq E$ we have

$$\mu(E') \le \mu(E) \ .$$

A vector bundle is said to be *stable* if the inequality is strict for every nonzero proper subbundle. A vector bundle E is *polystable* if it is isomorphic to a direct sum of stable bundles

$$E \simeq E_1 \oplus E_2 \oplus \cdots \oplus E_\ell$$

of the same slope $\mu(E_i)$, $i = 1, \ldots, \ell$. A vector bundle which is not semistable will be called *unstable* (Fig. 4.1).

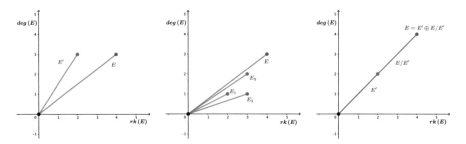

Fig. 4.1 By depicting slopes in a Cartesian diagram where the axis are the rank and the degree, we can visualize the stability notion. In the first figure, E is unstable because there exists $E' \subsetneq E$ with $\mu(E') > \mu(E)$. In the second figure, E is stable because for every $E_i \subsetneq E$ we have $\mu(E_i) < \mu(E)$, $i = 1, 2, 3$. In the third figure, E is strictly semistable because there exists $E' \subsetneq E$ with $\mu(E') = \mu(E)$; moreover, E is polystable because $E = E' \oplus E/E'$, with $\mu(E') = \mu(E/E') = \mu(E)$

Recall that the degree, and therefore, the notion of stability, does not depend on the polarization $O_X(1)$ of the curve X. This is not true for higher dimensional varieties X.

Let m be an integer. Recall from Sect. 2.3.1 that a vector bundle E over X is *m-regular* if $H^i(X, E(m-i)) = 0$, $\forall i > 0$. If E is m-regular, then the twist $E(m) := E \otimes O_X(m)$ is generated by global sections (i.e., the evaluation map is surjective). By Theorem 2.2, for each vector bundle E, there exists an integer m (depending on E) such that E is m-regular.

It can be proved (see [70, 80] for curves and [61] in higher dimension) that all semistable vector bundles of rank r and degree d are bounded, meaning that they are parametrized by a finite dimensional scheme whose ranks and degrees of their subbundles and quotients are bounded. Then it can be shown that, for a bounded family, we can choose the integer m uniformly in Theorem 2.2.

With this choice of m, the dimension of the space of global sections of the twisted bundle $E(m)$ is, by Riemann-Roch theorem,

$$h^0(X, E(m)) = \chi(E(m)) = \deg(E \otimes O_X(m)) + r(1-g) = d + rm + r(1-g) =: n \,,$$

which is a linear polynomial in m. Let V be an n-dimensional complex vector space and choose an isomorphism $\alpha : V \simeq H^0(X, E(m))$. By composing with the evaluation map, we obtain a surjection

$$V \otimes O_X \simeq H^0(X, E(m)) \otimes O_X \twoheadrightarrow E(m) \,.$$

Taking global sections, we get a homomorphism between vector spaces

$$H^0(X, V \otimes O_X) \longrightarrow H^0(X, E(m))$$

and, taking the r-exterior power with $r = \mathrm{rk}(E)$, obtain

$$\wedge^r V \longrightarrow \wedge^r H^0(X, E(m)) \longrightarrow H^0\left(X, \wedge^r(E(m))\right) =$$

$$H^0\left(X, \wedge^r E \otimes O_X(rm)\right) \simeq H^0(X, L \otimes O_X(rm)) =: W \,,$$

where the isomorphism comes from

$$\beta : \wedge^r E = \det(E) \simeq L$$

and two of these isomorphisms β differ by a scalar. Therefore, a point (E, α) provides an element in $\mathrm{Hom}(\wedge^r V, W)$ and, because of the choice of β, a well-defined element in $\mathbb{P}(\mathrm{Hom}(\wedge^r V, W))$. Then we have all semistable bundles, together with a choice of an isomorphism α, inside the so-called *Quot-scheme* (or scheme of quotients) defined by Grothendieck [34]:

$$Z := \{(E, \alpha)\} \subset \text{Quot}_{X,r,L,m} \hookrightarrow \mathbb{P}\left(\text{Hom}(\wedge^r V, W)\right) \ ,$$

parametrizing quotients $q : V \otimes O_X(-m) \twoheadrightarrow E$ corresponding in the projective space to linear maps $[q' : \wedge^r V \rightarrow W]$. To obtain a moduli space, we will take the GIT quotient of this subvariety Z by the action encoding the changes of the isomorphism α; this is the action of $\text{GL}(V) = \text{GL}(n, \mathbb{C})$ although, because of the projectivity, it is enough to take the quotient by $\text{SL}(n, \mathbb{C})$. The GIT quotient is taken with respect to the linearized action of $\text{SL}(n, \mathbb{C})$ on the ample line bundle of the projective embedding. By GIT results (c.f. Theorem 3.1), there exists a good quotient of the GIT semistable points Z^{ss}. Then, the moduli space of semistable holomorphic vector bundles over X of rank r and determinant L will be the GIT quotient

$$M(r, L) = Z^{ss} /\!/ \text{SL}(n, \mathbb{C}) \ .$$

Hilbert-Mumford criterion (c.f. Theorem 3.3) states that GIT stability can be checked by 1-parameter subgroups

$$\rho : \mathbb{C}^* \longrightarrow \text{SL}(n, \mathbb{C}) \ .$$

Let us sketch the computation of the weight of the action of ρ on a quotient $[q']$. The reader can see [71, Sections 4.6 and 5], [79, Section 2.2.4], or [53, Section 4.4] for the technical details of the construction.

Once fixed m, let V be a complex vector space of dimension $n := h^0(X, E(m))$. Let $\rho : \mathbb{C}^* \rightarrow \text{SL}(n, \mathbb{C})$ be a 1-parameter subgroup and let $[q'] \in \mathbb{P}(\text{Hom} \wedge^r V, W))$ be a point in the projective space corresponding to a quotient $q : V \otimes O_X(-m) \twoheadrightarrow E$. There exists a basis $\{v_1, v_2, \ldots, v_n\}$ of V such that

$$\rho(t) = \begin{pmatrix} t^{\rho_1} & 0 & \cdots & 0 \\ 0 & t^{\rho_2} & \cdots & 0 \\ \vdots & & \ddots & \vdots \\ 0 & 0 & \cdots & t^{\rho_n} \end{pmatrix}$$

acts on each v_i multiplying by t^{ρ_i}, and assume $\rho_1 \leq \rho_2 \leq \cdots \leq \rho_n$. Denote by

$$V_i := \mathcal{L}\{v_1, v_2, \ldots, v_i\} \quad \text{and} \quad V^i := V_i/V_{i-1} \ , \quad i = 1, 2, \ldots, n \ ,$$

the corresponding vector subspaces and quotients of V. Note that $\dim V_i = i$, $\dim V^i = 1$ and, since $\rho(t) \in \text{SL}(n, \mathbb{C})$, it is $\sum_{i=1}^{n} \rho_i = 0$.

Let $E_i := q\,(V_i \otimes O_X(-m))$ be the sheaf generated by the sections of V_i under the evaluation map and denote the successive quotients by $E^i := E_i/E_{i-1}$. It can be proved that, restricting to an open subset of the Quot-scheme, all E_i and E^i are locally free, hence they are vector bundles and we can also assume that all E_i are m-regular; therefore, $h^1\,(X, E_i(m)) = 0$ (see the aforementioned references). We denote $r_i := \mathrm{rk}(E_i)$ and $r^i := \mathrm{rk}(E^i)$, respectively, and note that $r^i = r_i - r_{i-1}$.

The action of $SL(n, \mathbb{C})$ on V yields an action on $\wedge^r V$ whose elements are linear combinations of $v_{i_1} \wedge \cdots \wedge v_{i_r}$, $i_1 < i_2 < \cdots < i_r$. Then, the weight of the action of ρ on $[q'] \in \mathbb{P}\,(\mathrm{Hom}\,\wedge^r V, W))$ is given by

$$\vartheta\,([q'], \rho) = \min\left\{\rho_{i_1} + \rho_{i_2} + \cdots + \rho_{i_r} : q'\left(v_{i_1} \wedge \cdots \wedge v_{i_r}\right) \neq 0\right\}$$

and the condition $q'\left(v_{i_1} \wedge \cdots \wedge v_{i_r}\right) \neq 0$ is equivalent to the image of the vectors v_{i_j}, $j = 1, \ldots, r$, under the evaluation map $q : V \otimes O_X(-m) \twoheadrightarrow E$, being linearly independent fiberwise in the generated subbundle of E. Then, since $r_i = \mathrm{rk}(E_i) = \mathrm{rk}\,(q\,(V_i \otimes O_X(-m)))$ and the exponents ρ_i are sorted increasingly, the minimum $\vartheta\,([q'], \rho)$ is achieved by getting as many times as possible the lowest exponent ρ_1, this is $r_1 = r^1$ times, then $r^2 = r_2 - r_1$ times the exponent ρ_2 and so on:

$$\vartheta\,([q'], \rho) = \sum_{i=1}^{n} \rho_i r^i .$$

On the one hand, using that $\displaystyle\sum_{i=1}^{n} \rho_i = 0$, it is an easy calculation to show that

$$\rho_n = \frac{1}{n} \sum_{i=1}^{n-1} i\,(\rho_{i+1} - \rho_i)$$

and, then, we get

$$\vartheta\,([q'], \rho) = \sum_{i=1}^{n} \rho_i\,(r_i - r_{i-1}) =$$

$$\rho_n r_n - \sum_{i=1}^{n-1} (\rho_{i+1} - \rho_i)\,r_i = \frac{1}{n} \sum_{i=1}^{n-1} (\rho_{i+1} - \rho_i)\,(ir - nr_i) .$$

If we have

$$\dim V r_i - \dim V_i r = n r_i - i r \geq 0 , \quad \text{for all } i,$$

it is $\vartheta\,([q'], \rho) \leq 0$ and the point $[q']$ is GIT semistable. On the other hand,

$$\vartheta\left([q'],\rho\right) = \sum_{i=1}^{n}\rho_i r^i = \sum_{i=1}^{n}\rho_i r^i - \frac{r}{n}\sum_{i=1}^{n}\rho_i =$$

$$\frac{1}{n}\sum_{i=1}^{n}\rho_i(r^i n - r) = \frac{1}{\dim V}\sum_{i=1}^{n}\rho_i\left(r^i\dim V - r\dim V^i\right).$$

Then, conversely, if there is a $V' \subset V$ generating E' of rank $r' := \mathrm{rk}(E')$ with

$$r'\dim V - r\dim V' < 0,$$

we can find a complement $V' \oplus V'' = V$ with V'' generating $E'' := E/E'$ of rank $r'' := \mathrm{rk}(E'') = r - r'$, and an adapted 1-parameter subgroup ρ acting as $\cdot t^{\rho_1}$ on V' and $\cdot t^{\rho_2}$ on V'', with $\rho_1 < 0$, $\rho_2 > 0$ and $\rho_1\dim V' + \rho_2\dim V'' = 0$, such that:

$$\vartheta\left([q'],\rho\right) = \frac{1}{\dim V}\left(\rho_1(r'\dim V - r\dim V') + \rho_2(r''\dim V - r\dim V'')\right) =$$

$$\frac{\rho_1}{\dim V}\left(r'\dim V - r\dim V'\right) - \frac{\rho_1\dim V'}{\dim V\dim V''}\left((r-r')\dim V - r(\dim V - \dim V')\right) =$$

$$\frac{\rho_1}{\dim V''}\left(r'\dim V - r\dim V'\right) > 0$$

because $\rho_1 < 0$; hence $[q']$ is GIT unstable. Therefore, we can conclude that E is semistable (resp. stable), if and only if,

$$r\dim V' \underset{(<)}{\leq} r'\dim V,$$

for all $V' \subsetneq V$, E' being the subbundle generated by V' and $r' = \mathrm{rk}(E')$.

Recall that $\dim V = h^0(X, E(m)) = \chi(E(m))$ because of the m-regularity of E, and note that $\dim V' < h^0(X, E'(m))$, the generated bundle $E'(m)$ having more sections in general. Then, if the point $[q']$ associated to (E, α) were a GIT unstable point, there would exists a subspace $V' \subset V$ generating the bundle $E'(m)$ such that

$$\frac{h^0(X, E'(m))}{r'} > \frac{\dim V'}{r'} > \frac{\dim V}{r} = \frac{h^0(X, E(m))}{r}.$$

Since we have, by Riemann-Roch theorem (c.f. Theorem 2.3):

$$\frac{h^0(X, E(m))}{r} = \frac{\chi(E(m))}{r} = \frac{d + rm + r(1 - g)}{r} =$$

$$\frac{d}{r} + m + (1 - g) = \mu(E) + m + (1 - g),$$

and similarly, for $E'(m)$:

$$\frac{h^0\left(X, E'(m)\right)}{r} = \mu(E') + m + (1 - g) ,$$

the inequality turns out to be

$$\mu(E') > \mu(E) ,$$

recovering the definition of unstability for vector bundles. This shows that an unstable vector bundle yields a GIT unstable point and vice versa, semistable bundles correspond to GIT semistable points (c.f. [71, Theorem 5.6], [79, Proposition 2.2.4.17], or [53, Lemma 4.4.5]).

Therefore, Z^{ss} matches, indeed, the semistable bundles that we want to classify, and the GIT quotient

$$\mathcal{M}(r, L) = Z^{ss} /\!\!/ \mathrm{SL}(n, \mathbb{C})$$

is the moduli space of semistable holomorphic vector bundles of rank r and determinant L over X. As GIT says, this moduli space is a good quotient where points correspond to S-equivalence classes of semistable vector bundles, each class containing a unique closed orbit, being the polystable representative. The stable bundles

$$\mathcal{M}_s(r, L) = Z^s / \mathrm{SL}(n, \mathbb{C})$$

yield a geometric quotient which is a quasi-projective variety.

If we do not fix the determinant of E to be a line bundle L, we can also perform this construction to obtain a moduli space $\mathcal{M}(r, d)$, which fibers over the component $\mathrm{Pic}^d(X) \subset \mathrm{Pic}(X)$ of the Picard group of X containing the line bundles of degree d.

4.2   Harder-Narasimhan Filtration

Unstable vector bundles fall outside the solution to the moduli problem explained in Sect. 4.1. However, to each unstable bundle, there is attached a canonical filtration called the *Harder-Narasimhan filtration* (see [37] and [53, Section 1.3]) which exhibits unstable bundles through successive quotients of semistable ones.

Theorem 4.1 *Let E be a vector bundle of rank r. There exists a unique filtration, called the* Harder-Narasimhan filtration *for E:*

$$0 = E_0 \subsetneq E_1 \subsetneq E_2 \subsetneq \cdots \subsetneq E_{t-1} \subsetneq E_t \subsetneq E_{t+1} = E$$

Satisfying the following:

1. The slopes of the quotients $E^i := E_i/E_{i-1}$ verify

$$\mu(E^1) > \mu(E^2) > \cdots > \mu(E^t) > \mu(E^{t+1}) .$$

2. The quotients E^i, for every $i = 1, \ldots, t + 1$, are semistable.

As we discussed in the construction of the moduli space, unstable bundles have the biggest possible automorphism group, coming from the different semistable blocks of their Harder-Narasimhan filtration and the homomorphisms between them. On the other hand, stable bundles are much easier because they are *simple* objects whose automorphisms are just the scalars. In between these two, semistable but non-stable bundles can be studied by means of their Jordan-Hölder filtration, whose graded object captures this automorphism group. This leads us to the notion of S-equivalence, the property to identify semistable points in the moduli space.

Theorem 4.2 *Let E be a semistable vector bundle of rank r. There exists a (nonunique in general) filtration, called the* Jordan-Hölder filtration *for E:*

$$0 = E_0 \subsetneqq E_1 \subsetneqq E_2 \subsetneqq \cdots \subsetneqq E_{\ell-1} \subsetneqq E_\ell \subsetneqq E_{\ell+1} = E$$

Satisfying the following:

1. The slopes of the quotients $E^i := E_i/E_{i-1}$ are equal:

$$\mu(E^1) = \mu(E^2) = \cdots = \mu(E^\ell) = \mu(E^{\ell+1}) .$$

2. The quotients E^i, for every $i = 1, \ldots, \ell + 1$, are stable.

The filtration is unique in the sense that the graded objects $\mathrm{gr}(E) \simeq \bigoplus_{i=1}^{\ell+1} E^i$ *of two different Jordan-Hölder filtrations of E are isomorphic.*

Definition 4.2 Let E and E' be vector bundles such that the graded objects of any of their Jordan-Hölder filtrations are isomorphic, i.e., $\mathrm{gr}(E) \simeq \mathrm{gr}(E')$. Then E and E' are said to be S-*equivalent*.

It happens that, if E and E' are S-equivalent bundles, their orbits in the GIT construction of the moduli space are S-equivalent in the sense of GIT, and they correspond to the same point in the moduli space. Vector bundles E which are isomorphic to the graded object of any of their Jordan-Hölder filtrations, this is

$$E \simeq E^1 \oplus E^2 \oplus \cdots \oplus E^{\ell+1}$$

where all quotient factors E^i are stable bundles of the same slope $\mu(E^i)$, are precisely the polystable vector bundles corresponding to GIT polystable orbits in

Fig. 4.2 Harder-Narasimhan polygon corresponding to a vector bundle E

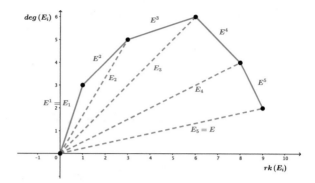

the GIT construction, the only closed orbits within each S-equivalence class. This completes the classification of points in the moduli problem of bundles.

From Theorem 4.1, we can define (following Shatz [83]) the *Harder-Narasimhan polygon* as the polygon in the first quadrant of the plane with vertices $(\mathrm{rk}(E_i), \deg(E_i))$ for $i = 0, \ldots, t + 1$ (see the points in Fig. 4.2). The slope of the line joining $(\mathrm{rk}(E_{i-1}), \deg(E_{i-1}))$, and $(\mathrm{rk}(E_i), \deg(E_i))$ is the slope of the quotient $\mu\left(E_i/E_{i-1}\right) = \mu(E^i)$ (line segments in Fig. 4.2). Slopes $\mu(E_i)$ of each factor in the filtration are represented by dashed lines in Fig. 4.2. The *Harder-Narasimhan type* of E is defined as the vector in $\mathbb{Q}^r$:

$$\vec{\mu}(E) := \left(\mu(E^1), \ldots, \mu(E^1), \mu(E^2), \ldots, \mu(E^2), \ldots, \mu(E^{t+1}), \ldots, \mu(E^{t+1})\right)$$

where the slope of each E^j appears $\mathrm{rk}\left(E^j\right)$ times.

In Fig. 4.2, it is represented the Harder-Narasimhan polygon corresponding to a vector bundle E of slope $\mu(E) = \frac{2}{9}$. Its Harder-Narasimhan type is

$$\vec{\mu}(E) = \left(\mu(E^1), \mu(E^2), \mu(E^2), \mu(E^3), \mu(E^3), \mu(E^3), \mu(E^4), \mu(E^4), \mu(E^5)\right)$$

$$= \left(3, 1, 1, \frac{1}{3}, \frac{1}{3}, \frac{1}{3}, -1, -1, -2\right),$$

where we observe the decreasing pattern in the slopes of the successive quotients of the Harder-Narasimhan filtration.

First condition in Theorem 4.1 states that $\mu\left(E^i\right) > \mu\left(E^{i+1}\right)$ for $i = 1, \ldots, t$, being equivalent to say that the Harder-Narasimhan polygon is convex, which is intuitively clear from the definition of the Harder-Narasimhan polygon (see [83, Proposition 5]). Second condition assures that all quotients are semistable, which can be rephrased as the Harder-Narasimhan polygon being the convex envelope of every refinement. In other words, if there is an unstable quotient, it has a destabilizing subbundle of greater slope which, therefore, will be depicted above the polygon.

4.3   Other Constructions of the Moduli Space of Vector Bundles

Most of the material of this book is concerned with classical algebraic methods of classifying objects in a moduli space, mainly geometric invariant theory. However, the story of classification problems with vector bundles is much richer and is addressed from many intertwined points of view.

Complementary works of Narasimhan-Seshadri and Mumford's GIT provide the first definition of stable bundle and a construction of the moduli space of holomorphic vector bundles. In the first case [70], it is done by relating them to irreducible unitary representations of the (universal central extension of the) fundamental group of a Riemann surface. Then, Mumford [66] and Seshadri [80] develop the theory to compactify this moduli space by adding the S-equivalence classes of semistable objects.

Previously, Grothendieck [33] had classified all vector bundles over the projective line $\mathbb{P}^1_{\mathbb{C}}$ as direct sums of line bundles (then the moduli space of semistable bundles happens to be a point or empty, for each degree and rank), and Atiyah [4] shows that the indecomposable bundles over a genus one projective curve (also called an *elliptic curve*) are in correspondence with the points of the curve. When the genus is greater than two, more subtleties show up.

Then, from the analytical point of view, we can construct the same moduli space of holomorphic vector bundles as the problem of classifying holomorphic structures in a bundle with just a differentiable structure. Coming from representation theory, we will obtain another homeomorphic space with the same topology, however not recovering the complex geometry. The book [92] by Wells with its Appendix by García-Prada covers this material in detail.

4.3.1   Analytical Construction of the Moduli Space of Vector Bundles

Let X be a smooth complex projective variety of dimension n, equivalent to a compact complex analytic manifold. Let TX be the tangent bundle of X, whose fibers are the tangent spaces TX_x to each $x \in X$, and let $TX^\vee$ be its cotangent bundle. Denote by

$$\wedge TX = \bigoplus_{r=0}^{n} \wedge^r TX \quad \text{and} \quad \wedge TX^\vee = \bigoplus_{r=0}^{n} \wedge^r TX^\vee$$

the *exterior algebra bundles*, which are vector bundles with fibers isomorphic to the anti-symmetric tensor product of degree r of the tangent TX_x and cotangent $TX_x^\vee$ vector spaces.

Recall the notation from Sect. 2.2.1 where $\mathcal{E}_X^k$ is the sheaf of smooth differential forms of degree k on X which, on each open subset $U \subset X$, takes values in $\wedge^k TX^\vee$. Denote by $\mathcal{E}_X^0 := \mathcal{E}_X$ the sheaf of smooth functions on X and define the *exterior derivative* as a morphism of sheaves which, on each $U \subset X$, is a map

$$d : \mathcal{E}_X^k(U) \to \mathcal{E}_X^{k+1}(U)$$

that generalizes the differential of functions. Indeed, for $f \in \mathcal{E}_X^0(U)$, it is

$$df = \sum_{j=1}^{n} \frac{\partial f}{\partial x_j} dx_j + \sum_{j=1}^{n} \frac{\partial f}{\partial y_j} dy_j = \sum_{j=1}^{n} \frac{\partial f}{\partial z_j} dz_j + \sum_{j=1}^{n} \frac{\partial f}{\partial \bar{z}_j} d\bar{z}_j \,,$$

where $z_1 = x_1 + iy_1, \ldots, z_n = x_n + iy_n$ are complex coordinates of X and $\bar{z}_j = x_j - iy_j$.

The complex structure of X gives rise to a complex structure on TX, giving the decomposition

$$TX = TX^{1,0} \oplus TX^{0,1}$$

into bundles whose fibers are $\pm i$-eigenspaces. Then, by duality we have

$$TX^\vee = TX^{\vee\, 1,0} \oplus TX^{\vee\, 0,1}.$$

Taking exterior powers of these bundles, let $\wedge^{p,q} TX^\vee$ be the bundle of smooth differential complex-valued (p, q)-forms on X, whose sheaf of sections is denoted by $\mathcal{E}_X^{p,q}$. Locally, these sections are given by the wedge product of p holomorphic differentials $dz_j = dx_j + idy_j$ and q anti-holomorphic differentials $d\bar{z}_j = dx_j - idy_j$.

We have the decomposition

$$\mathcal{E}_X^k = \sum_{p+q=k} \mathcal{E}_X^{p,q}$$

and, hence, the exterior derivative d restricts to

$$d : \mathcal{E}_X^{p,q} \longrightarrow \sum_{s+t=p+q+1} \mathcal{E}_X^{s,t}$$

which breaks into the ∂ and $\bar{\partial}$ operators by restricting to the corresponding subspaces:

$$\partial : \mathcal{E}_X^{p,q} \longrightarrow \mathcal{E}_X^{p+1,q} \ , \quad \bar{\partial} : \mathcal{E}_X^{p,q} \longrightarrow \mathcal{E}_X^{p,q+1} \ .$$

It happens that $d^2 := d \circ d = 0$ and $\bar{\partial}^2 = 0$ (a proof of this is Grothendieck's generalization of Poincaré lemma for the operator $\bar{\partial}$, see [32, Chapter 0, Section 2, p.25]), then we obtain a resolution of sheaves

$$0 \longrightarrow \Omega_X^p \overset{\iota}{\longrightarrow} \mathcal{E}_X^{p,0} \overset{\bar{\partial}}{\longrightarrow} \mathcal{E}_X^{p,1} \overset{\bar{\partial}}{\longrightarrow} \cdots \overset{\bar{\partial}}{\longrightarrow} \mathcal{E}_X^{p,n} \longrightarrow 0 \ ,$$

where ι is the inclusion of Ω_X^p, the sheaf of holomorphic differential forms of degree p, which is the kernel of

$$\bar{\partial} : \mathcal{E}_X^{p,0} \longrightarrow \mathcal{E}_X^{p,1} \ .$$

We can extend these definitions to sheaves of forms with values in a vector bundle. Let $\mathbb{E} \to X$ be a smooth complex vector bundle, whose fibers are complex vector spaces, but its transition functions are complex differentiable not necessarily holomorphic. Identify $\mathbb{E}$ with the sheaf of smooth sections with values in complex vector spaces and define

$$\mathcal{E}_X^k(\mathbb{E}) := \mathcal{E}_X^k \otimes \mathbb{E} \ ,$$

the sheaves of smooth $\mathbb{R}$-linear differential forms on X of degree k with values in (the smooth sections of) $\mathbb{E}$. This sheaf is the same as the sheaf of smooth sections of the smooth complex vector bundle $\wedge^k(TX^\vee) \otimes_{\mathbb{R}} \mathbb{E}$. Using the complex structure of X, define also

$$\mathcal{E}_X^{p,q}(\mathbb{E}) := \mathcal{E}_X^{p,q} \otimes \mathbb{E} \ ,$$

sheaves of $\mathbb{R}$-linear differential (p,q)-forms with values in $\mathbb{E}$.

Locally, vector bundles are trivial, then differentiating functions follows traditional calculus rules, once fixed some coordinates. When these functions take values in a bundle, to measure the infinitesimal variation of such functions, we need to connect the fibers mathematically: this leads to the notion of *connection*.

Definition 4.3 Let $\mathbb{E}$ be a smooth complex vector bundle on a compact Riemann surface X. A *connection* on $\mathbb{E} \to X$ is a first-order differential operator

$$\nabla : \mathcal{E}_X^0(\mathbb{E}) \longrightarrow \mathcal{E}_X^1(\mathbb{E})$$

satisfying Leibniz rule $\nabla(fs) = f\nabla(s) + df \otimes s$, where $f \in \mathcal{E}_X^0$ is a smooth function on X, $s \in \mathcal{E}_X^0(\mathbb{E})$ is a smooth section of $\mathbb{E}$, and d is the exterior derivative.

A connection on a compact Riemann surface X splits into

$$\nabla = \nabla^{1,0} \oplus \nabla^{0,1} : \mathcal{E}_X^0(\mathbb{E}) \longrightarrow \mathcal{E}_X^{1,0}(\mathbb{E}) \oplus \mathcal{E}_X^{0,1}(\mathbb{E})$$

where the $(0, 1)$ part satisfies the Leibniz rule $\nabla^{0,1}(fs) = f\nabla^{0,1}(s) + \bar{\partial}f \otimes s$. Let us define a *Dolbeault operator* as a differential operator

$$\bar{\partial}_\mathbb{E} : \mathcal{E}_X^0(\mathbb{E}) \longrightarrow \mathcal{E}_X^{0,1}(\mathbb{E})$$

satisfying the same Leibniz rule as $\nabla^{0,1}$.

Now, let E be a holomorphic bundle over an n-dimensional smooth complex projective variety X and define

$$\Omega_X^p(E) := \Omega_X^p \otimes E$$

to be the sheaf of holomorphic differential forms of degree p with values in (the holomorphic sections of) E, which is equal to the sheaf of holomorphic sections of $\wedge^k (TX^\vee) \otimes_\mathbb{C} E$. Similarly,

$$\mathcal{E}_X^k(E) := \mathcal{E}_X^k \otimes E \quad \text{and} \quad \mathcal{E}_X^{p,q}(E) := \mathcal{E}_X^{p,q} \otimes E$$

are the sheaves of smooth sections of the bundles $\wedge^k (TX^\vee) \otimes_\mathbb{C} E$ and $\wedge^{p,q} (TX^\vee) \otimes_\mathbb{C} E$, respectively. Extending the definition of $\bar{\partial}$ to $\bar{\partial}_E := \bar{\partial} \otimes 1$, a fine resolution of sheaves can be stated in this situation as follows:

$$0 \longrightarrow \Omega_X^p(E) \overset{i}{\longrightarrow} \mathcal{E}_X^{p,0}(E) \overset{\bar{\partial}_E}{\longrightarrow} \mathcal{E}_X^{p,1}(E) \overset{\bar{\partial}_E}{\longrightarrow} \cdots \overset{\bar{\partial}_E}{\longrightarrow} \mathcal{E}_X^{p,n}(E) \longrightarrow 0 .$$

Similar to De Rham's Theorem in Sect. 2.2.1, we can calculate the cohomology groups of $\Omega_X^p(E)$ by a resolution of smooth differential (p, q)-forms with values in E, generalizing Dolbeault's theorem.

Theorem 4.3 ([20]) *Given a holomorphic bundle E over a compact complex analytic manifold X, the cohomology with coefficients in the sheaf of holomorphic differential p-forms with values in E can be computed as*

$$H^q(X, \Omega_X^p(E)) = \frac{\mathrm{Ker}\left(\mathcal{E}_X^{p,q}(E)(X) \overset{\bar{\partial}_E}{\longrightarrow} \mathcal{E}_X^{p,q+1}(E)(X)\right)}{\mathrm{Im}\left(\mathcal{E}_X^{p,q-1}(E)(X) \overset{\bar{\partial}_E}{\longrightarrow} \mathcal{E}_X^{p,q}(E)(X)\right)} .$$

For $p = 0$, the space $\Omega_X^0(E)$ represents the holomorphic differential 0-forms with values in E, which are precisely those local sections of the bundle E which are not only differentiable but, indeed, holomorphic, i.e., satisfying the Cauchy-Riemann equations; therefore, we identify $\Omega_X^0(E) = E$. Then, the cohomology of the holomorphic vector bundle $H^q(X, E)$ can also be computed with the resolution given by this complex, what is called *Dolbeault cohomology*:

$$H^q(X, E) = \frac{\text{Ker}\left(\mathcal{E}_X^{0,q}(E)(X) \xrightarrow{\bar{\partial}_E} \mathcal{E}_X^{0,q+1}(E)(X) \right)}{\text{Im}\left(\mathcal{E}_X^{0,q-1}(E)(X) \xrightarrow{\bar{\partial}_E} \mathcal{E}_X^{0,q}(E)(X) \right)} .$$

If we particularize the resolution complex for a compact Riemann surface X, we see that

$$\mathcal{E}_X^2 = 0 \quad \text{and} \quad \mathcal{E}_X^{0,2} = \mathcal{E}_X^{2,0} = 0 ,$$

and the condition $\bar{\partial}_E^2 = 0$ is automatically satisfied. Then, a holomorphic vector bundle E over a compact Riemann surface yields a Dolbeault operator $\bar{\partial}_E$ with $\bar{\partial}_E^2 = 0$. Conversely, it can be shown that a Dolbeault operator satisfying $\bar{\partial}_E^2 = 0$ provides local solutions for the equation $\bar{\partial}_E(s) = 0, s \in \mathcal{E}_X^0$, and then a holomorphic structure on $\mathbb{E}$.

Getting everything together, a holomorphic vector bundle E is determined by a smooth complex bundle $\mathbb{E}$ plus a Dolbeault operator $\bar{\partial}_E$:

$$E = (\mathbb{E}, \bar{\partial}_E) ,$$

where the group of automorphisms of $\mathbb{E}$, called the *complex gauge group* $\mathcal{G}^{\mathbb{C}}$, acts by conjugation:

$$g \cdot \bar{\partial}_E = g \bar{\partial}_E g^{-1} , \quad g \in \mathcal{G}^{\mathbb{C}} .$$

Fixing a smooth complex vector bundle $\mathbb{E}$ over X, the rank r and the degree d are also fixed because these are topological invariants. Denote by $\mathcal{A}^{0,1}(\mathbb{E})$ the set of holomorphic structures $\bar{\partial}_E$ with rank r and degree d. Then, the quotient $\mathcal{A}^{0,1}(\mathbb{E})/\mathcal{G}^{\mathbb{C}}$ by the complex gauge group can be identified with the set of isomorphism classes of holomorphic vector bundles of rank r and degree d.

Once stated this, we can talk about smooth complex subbundles $\mathbb{F} \subset \mathbb{E}$ and holomorphic subbundles $F \subset E$ given by restrictions of the holomorphic structure $\bar{\partial}_E|_F$, for which the slope

$$\mu(F) = \frac{\deg(F)}{\mathrm{rk}(F)}$$

can be defined and as well as the notion of stability (see Definition 4.1). Nevertheless, the quotient $\mathcal{A}^{0,1}(\mathbb{E})/\mathcal{G}^{\mathbb{C}}$ is not Hausdorff, so in order to have an interesting quotient, we need to impose the stability condition. Define by

$$\mathcal{A}_s^{0,1}(\mathbb{E}) = \left\{\bar{\partial}_E \in \mathcal{A}^{0,1}(\mathbb{E}) : E = (\mathbb{E}, \bar{\partial}_E) \text{ is stable}\right\}$$

and, similarly, semistable and polystable holomorphic structures $\mathcal{A}_{ss}^{0,1}(\mathbb{E})$, $\mathcal{A}_{ps}^{0,1}(\mathbb{E})$, respectively. Then we obtain the analytical (or Dolbeault) moduli space as the gauge theoretical quotient of the polystable structures.

Theorem 4.4 *The moduli space of stable vector bundles E of rank r and degree d over a compact Riemann surface X of genus g, defined as the gauge quotient*

$$\mathcal{N}_s(r, d) = \mathcal{A}_s^{0,1}(\mathbb{E})/\mathcal{G}^{\mathbb{C}},$$

is a smooth complex analytic manifold.

The space $\mathcal{A}^{0,1}(\mathbb{E})$ is an affine space modeled on the vector space $\mathcal{E}_X^{0,1}(\mathrm{End}\,\mathbb{E})$. Then, the tangent space to $\mathcal{N}_s(r, d)$ at a point $(E, \bar{\partial}_E)$ is this vector space; therefore, the dimension of the moduli space is the dimension of $\mathcal{E}_X^{0,1}(\mathrm{End}\,\mathbb{E})$ which is equal to $h^1(X, \mathrm{End}\,E)$. The vector bundle $\mathrm{End}\,E$ has rank r^2 and degree 0 (note that $\mathrm{End}\,E = \mathrm{Hom}(E, E) = E^{\vee} \otimes E$), then by Riemann-Roch:

$$\chi(\mathrm{End}\,E) = h^0(X, \mathrm{End}\,E) - h^1(X, \mathrm{End}\,E) = -r^2(g-1)\,.$$

And stable vector bundles are simple, i.e., $H^0(X, \mathrm{End}\,E) = \mathbb{C}$, then $h^0(X, \mathrm{End}\,E) = 1$ and, hence, the dimension of the moduli space of stable bundles is

$$\dim \mathcal{N}_s(r, d) = h^1(X, \mathrm{End}\,E) = 1 + r^2(g-1).$$

Using Mumford's theory [66], we can compactify this moduli space of stable bundles by using semistable ones. It happens that on each S-equivalence class (by the action of the gauge group in the sense of GIT), there is a unique polystable holomorphic structure, then:

Theorem 4.5 *The moduli space of semistable vector bundles E of rank r and degree d over a compact Riemann surface X is the complex projective variety given as the quotient of the semistable or the polystable structures by the complex gauge group*

$$\mathcal{N}(r, d) = \mathcal{A}_{ss}^{0,1}(\mathbb{E}) /\!/ \mathcal{G}^{\mathbb{C}} = \mathcal{A}_{ps}^{0,1}(\mathbb{E})/\mathcal{G}^{\mathbb{C}}.$$

It contains the moduli of stable bundles $N_s(r, d) \subset N(r, d)$ *as a quasi-projective variety.*

When the rank and degree are coprime, there are semistable bundles which are not stable and we can simplify the previous results (c.f. [70]):

Theorem 4.6 *If* $\mathrm{GCD}(r, d) = 1$, *then* $\mathcal{A}_{ps}^{0,1}(\mathbb{E}) = \mathcal{A}_s^{0,1}(\mathbb{E})$ *and* $N(r, d) = N_s(r, d)$ *is a smooth complex projective variety.*

This analytical moduli space $N(r, d)$ is equivalent to the one defined in Sect. 4.1. The space $N(r, d)$ maps:

$$N(r, d) \longrightarrow \mathrm{Pic}^d(X) \subset \mathrm{Pic}(X)$$

over the component of the Picard group of degree d line bundles over X. The fibers of this map are vector bundles with determinant isomorphic to a fixed line bundle L of degree d, then the fibers are isomorphic to the moduli spaces $M(r, L)$ of holomorphic vector bundles of rank r and determinant isomorphic to a fixed line bundle L, explored in Sect. 4.1, and the total moduli spaces are isomorphic

$$N(r, d) \simeq M(r, d).$$

4.3.2　Moduli Space of Representations of the Fundamental Group

In this section, we want to approach the classification problem from the point of view of representations, which will be associated to flat connections on vector bundles.

Let $\mathbb{E}$ be a smooth complex vector bundle on a compact Riemann surface X and let

$$\nabla : \mathcal{E}_X^0(\mathbb{E}) \longrightarrow \mathcal{E}_X^1(\mathbb{E})$$

be a connection on $\mathbb{E}$. Once we have a way to take derivatives of functions along the fibers of $\mathbb{E}$, it makes sense to define a section $s \in \mathcal{E}_X^0(\mathbb{E})$ to be *parallel* if $\nabla s = 0$. Given a curve $\gamma(t)$ in X such that $\gamma(0) = x_0 \in X$ and a vector $v \in \mathbb{E}_{x_0}$, there is a unique parallel section $s \in \mathcal{E}_X(\mathbb{E})$ such that $s(\gamma(0)) = s(x_0) = v$, called the *parallel displacement* of v along γ and taking values $s(\gamma(t)) \in \mathbb{E}_{\gamma(t)}$. Applying twice the operator ∇ we obtain the *curvature* of a connection, $\nabla^2 = \nabla \circ \nabla = \Theta_\nabla$:

$$\Theta_\nabla : \mathcal{E}_X^0(\mathbb{E}) \longrightarrow \mathcal{E}_X^2(\mathbb{E})$$

which represents, with respect to this idea of parallel displacement, how curved the total space of $\mathbb{E}$ is. A connection such that $\Theta_\nabla = 0$ is called *flat*.

Let $\pi_1(X, x_0)$ be the fundamental group of X. Its elements are classes of closed loops

$$\gamma : [0, 1] \longrightarrow X , \quad \gamma(0) = \gamma(1) = x_0 ,$$

by homotopy equivalence, and it is independent of x_0 when X is connected. For a flat connection ∇ on $\mathbb{E}$, the parallel displacement produces a map

$$\gamma \in \pi_1(X, x_0) \longmapsto [v = s(\gamma(0)) \mapsto s(\gamma(1))] \in \mathrm{GL}(\mathbb{E}_{x_0})$$

defining a *representation of the fundamental group* into the general linear group

$$\xi : \pi_1(X, x_0) \longrightarrow \mathrm{GL}(r, \mathbb{C}) ,$$

whose image is the *holonomy group* of the connection ∇, where r is the rank of $\mathbb{E}$.
Conversely, any such representation defines

$$\mathbb{E} = \widetilde{X} \times_\xi \mathbb{C}^r := \widetilde{X} \times \mathbb{C}^r / \sim$$

where $\widetilde{X} \to X$ is the universal covering of X. Given a loop γ with $\gamma(0) = \gamma(1) = x$, let $\widetilde{\gamma}$ be a lift to $\widetilde{X} \to X$ such that $\widetilde{\gamma}(0)$ and $\widetilde{\gamma}(1)$ lie over x. The equivalence relation is given by

$$(x, w) \longmapsto (\widetilde{\gamma}(1), \xi(\gamma)w) .$$

It happens that flat connections exist only for smooth bundles $\mathbb{E}$ whose first Chern class is $c_1(\mathbb{E}) = 0$, then they correspond to degree 0 bundles. To obtain bundles with different degrees, we need to consider *projectively flat connections* which are connections whose holonomy maps are defined as

$$\widetilde{\xi} : \pi_1(X, x_0) \longrightarrow \mathrm{PGL}(r, \mathbb{C})$$

and lifted to representations

$$\xi : \Gamma \longrightarrow \mathrm{GL}(r, \mathbb{C}) ,$$

where Γ is a *universal central extension* of the fundamental group:

$$0 \longrightarrow \mathbb{Z} \longrightarrow \Gamma \longrightarrow \pi_1(X, x_0) \longrightarrow 0 .$$

Thus, let $\mathbb{E}$ be a smooth complex vector bundle of rank r and degree d on a compact Riemann surface X. Consider a *hermitian metric* h on $\mathbb{E}$, i.e., an hermitian inner product $\langle , \rangle_x$ on each fiber $\mathbb{E}_x$. It can be shown that the holonomy of a connection ∇ which is compatible with h lives in $\mathrm{U}(r) \subset \mathrm{GL}(r, \mathbb{C})$, because unitary matrices represent base changes in hermitian structures. Then, define

$$\mathrm{Hom}\,(\Gamma, \mathrm{GL}(r, \mathbb{C})) := \{\xi : \Gamma \longrightarrow \mathrm{GL}(r, \mathbb{C})\} \quad \text{and}$$

$$\mathrm{Hom}^d\,(\Gamma, \mathrm{GL}(r, \mathbb{C})) := \{\xi : \Gamma \longrightarrow \mathrm{GL}(r, \mathbb{C}) : c_1(\mathbb{E}) = d\},$$

where $c_1(\mathbb{E})$ is the first Chern class of the vector bundle associated to a representation following the previous construction. The group $\mathrm{GL}(r, \mathbb{C})$ acts by conjugation on representations:

$$(g \cdot \xi)(\gamma) = g\xi(\gamma)g^{-1},$$

where $g \in \mathrm{GL}(r, \mathbb{C})$, $\xi \in \mathrm{Hom}\,(\Gamma, \mathrm{GL}(r, \mathbb{C}))$ and $\gamma \in \Gamma$.

Finally, define the affine GIT quotient

$$\mathcal{R}(r, d) = \mathrm{Hom}^d\,(\Gamma, \mathrm{U}(r)) /\!/ \mathrm{U}(r)$$

to be the *moduli space of representations* of degree d of a universal central extension Γ of the fundamental group of a compact Riemann surface X into the unitary group $\mathrm{U}(r)$, and let the geometric quotient

$$\mathcal{R}_{irr}(r, d) = \mathrm{Hom}^d_{irr}\,(\Gamma, \mathrm{U}(r)) / \mathrm{U}(r)$$

be the corresponding *moduli space of irreducible unitary representations*. Narasimhan and Seshadri show that these spaces correspond topologically to the moduli spaces of holomorphic vector bundles.

Theorem 4.7 ([70]) *There exist a homeomorphism between the moduli space of representations and the moduli space of semistable holomorphic vector bundles over a compact Riemann surface X:*

$$\mathcal{R}(r, d) \simeq \mathcal{N}(r, d) \simeq \mathcal{M}(r, d)$$

under which irreducible representations correspond to stable bundles:

$$\mathcal{R}_{irr}(r, d) \simeq \mathcal{N}_s(r, d) \simeq \mathcal{M}_s(r, d).$$

An additional ingredient to the formulation of last theorem can be stated in terms of connections on bundles and hermitian metrics, in the following way. It turns out that polystable holomorphic bundles are precisely those for which we can find a harmonic metric h in the total space of the bundle such that a canonical connection ∇_h (determined by h) satisfies an equation which is usually referred by saying that ∇ has *constant central curvature* (which is, in turn, equivalent to being a projective connection). Then, given a smooth

(continued)

complex vector bundle $\mathbb{E}$ over a Riemann surface X, and a hermitian metric h on the total space of $\mathbb{E}$, the moduli space of holomorphic vector bundles can also be derived as the space of connections of constant central curvature modded out by the gauge transformations. This approach is followed by Atiyah and Bott in [5] using Morse theory on the Yang-Mills flow, and Donaldson [18]. For a survey on the subject, we refer the reader to [92, Appendix, 2.2].

4.4 Moduli Space of Higgs Bundles

Higgs bundles were introduced by Hitchin in 1987 to study the Yang-Mills equations over Riemann surfaces. They represent the mathematical formulation of the scalar field that in the *Higgs mechanism* interacts with gauge bosons making them behave as if they carry mass. This theory intends to explain the symmetry break in particle physics.

Definition 4.4 A *Higgs bundle* over a compact Riemann surface X is a pair (E, φ) where E is a holomorphic bundle over X and $\varphi : E \to E \otimes \omega_X \in H^0(X, \text{End}(E) \otimes \omega_X)$ is the *Higgs field*, with ω_X the canonical bundle of X.

Similarly to holomorphic vector bundles, by introducing a stability condition, a moduli space of polystable Higgs bundles can be constructed yielding a complex algebraic variety. Original work [45] for rank 2 bundles shows the so-called *Hitchin-Kobayashi correspondence* where polystable Higgs bundles correspond to certain reductions of the structure group $\text{SL}(2, \mathbb{C})$ (in the language of metrics and connections), satisfying generalized Yang-Mills equations. The moduli space of Higgs bundles turns out to be homeomorphic to the Dolbeault moduli space classifying operators $\bar{\partial}$ in a smooth bundle, plus the data of the Higgs field, by gauge equivalence transformations. This proof is generalized by Simpson [85, 86] and García-Prada, Gothen, and Mundet i Riera [21] for higher rank Higgs bundles, and for other Lie groups G, leading to the notion of G-Higgs bundle.

The other fundamental piece in the Higgs bundle universe is its relationship with the representations $\xi : \pi_1(X) \to G$ of the fundamental group of X into G. Works of Donaldson [19], Uhlenbeck-Yau [90], Corlette [17], and Simpson [84, 85], among others, what is known as the non-abelian Hodge theory, provide a homeomorphism of the moduli space of Higgs bundles with the moduli of reductive representations, and extends the theorem of Narasimhan-Seshadri [70], adding the Higgs field on the bundle side, and the trip from the real compact form to the group G, generalizing the case $\text{U}(r) \subset \text{GL}(r, \mathbb{C})$. This way, it is completed a correspondence between three moduli spaces coming from algebra, geometry, topology, and physics.

4.4.1 Hitchin's Construction

Let $\mathbb{E}$ be a smooth complex vector bundle of rank r and degree d on a compact Riemann surface X. By Serre duality, we have

$$H^1(X, \text{End}(\mathbb{E}))^\vee \simeq H^0(X, \text{End}(\mathbb{E}) \otimes \omega_X) ,$$

then *Higgs fields* are elements of (the dual of) this first cohomology group of the endomorphism bundle. The moduli space of Higgs bundles can be derived from

$$\mathcal{H} := \left\{ (\bar{\partial}_E, \varphi) \in \mathcal{A}^{0,1}(\mathbb{E}) \times \mathcal{E}_X^{1,0}(\text{End}(\mathbb{E})) \right\}$$

modded out by the action

$$g \cdot (\bar{\partial}_E, \varphi) = \left(g\bar{\partial}_E g^{-1}, g\varphi g^{-1} \right) .$$

The work of Hitchin [45, 46] presents a construction resembling on an extension of the stability notion to Higgs bundles. In this case, the stability condition is analogous to the one for vector bundles, but checking just on subbundles preserved by φ.

Definition 4.5 Given a Higgs bundle (E, φ) over X, a subbundle $F \subset E$ is said to be φ-*invariant* if $\varphi(F) \subset F \otimes \omega_X$. A Higgs bundle is *semistable* if $\mu(F) \leq \mu(E)$ for any non-zero φ-invariant proper subbundle $F \subsetneq E$, *stable* if $\mu(F) < \mu(E)$ for any non-zero φ-invariant proper subbundle $F \subsetneq E$, *polystable* if it is the direct sum of stable φ-invariant subbundles of the same slope, and *unstable* if it is not semistable.

Therefore, a moduli space of stable (resp. polystable or S-equivalence classes of) Higgs bundles is the quotient of pairs $(\bar{\partial}_E, \varphi)$ defining a stable (resp. polystable) Higgs bundle (E, φ) by the complex gauge group:

$$\mathcal{M}_s^H(r, d) = \mathcal{H}_s/\mathcal{G}^{\mathbb{C}} , \quad \mathcal{M}^H(r, d) = \mathcal{H}_{ps}/\mathcal{G}^{\mathbb{C}} ,$$

where $\mathcal{H}_s, \mathcal{H}_{ps} \subset \mathcal{H}$ are the corresponding stable and polystable pairs.

Alternatively, Hitchin [46] describes the moduli space as the solutions to certain equations on metrics, called *Hitchin equations*. Let h be a hermitian metric on $\mathbb{E}$ and let C_h be the space of connections ∇ on $\mathbb{E}$ which are compatible with h. The seminal work [45] for the case $(r, d) = (2, 1)$ considers the space of pairs

$$(\nabla, \varphi) \in X := C_h \times \mathcal{E}_X^{1,0}(\text{End}(\mathbb{E}))$$

and the equations

$$\begin{cases} \Theta_\nabla + [\varphi, \varphi^*] = \delta \\ \\ \bar{\partial}_E \varphi = 0 \end{cases} \tag{4.1}$$

where Θ_∇ is the curvature of the connection ∇, φ^* is the adjoint of φ with respect to the hermitian metric h on $\mathbb{E}$, the bracket is given by $[\varphi, \varphi^*] = \varphi\varphi^* + \varphi^*\varphi$, and δ is the constant central curvature ($\delta = 0$ for $\deg(\mathbb{E}) = 0$). These equations are a reduction of the *Yang-Mills self-duality equations* (SDE) from $\mathbb{R}^4$ to $\mathbb{R}^2 \simeq \mathbb{C}$ [94].

It turns out that polystable Higgs bundles (E, φ) admit an hermitian metric h and a connection ∇ satisfying equations (4.1) and, conversely, Higgs bundles with a metric and a connection satisfying (4.1) are polystable. Therefore, the subset $X_0 \subset X$ of solutions to the equations (4.1) corresponds, up to the action of the complex gauge group $\mathcal{G}^{\mathbb{C}}$, to the moduli space of polystable Higgs bundles:

$$\mathcal{M}^{YM}(r, d) = X_0/\mathcal{G}^{\mathbb{C}} .$$

The two constructions are shown to yield homeomorphic moduli spaces for polystable Higgs bundles in [45]:

$$\mathcal{M}^{YM}(r, d) \simeq \mathcal{M}^H(r, d) ,$$

restricting to a homeomorphism between irreducible solutions X^* and stable Higgs bundles:

$$X^*/\mathcal{G}^{\mathbb{C}} = \mathcal{M}^{YM}_{irr}(r, d) \simeq \mathcal{M}^H_s(r, d) .$$

4.4.2 Higher Rank and Dimensional Higgs Bundles

For higher rank Higgs bundles with $r > 2$, it is Nitsure [72] who constructs the generalized moduli space $\mathcal{M}(r, d, L)$ of *semistable pairs*, which are pairs

$$(E, \varphi : E \to E \otimes L)$$

where E is a rank r and degree d holomorphic vector bundle over a smooth projective algebraic curve X, and L is a line bundle over X. It is a GIT construction generalizing the ideas of the moduli space of vector bundles in Sect. 4.1. The dimension of these moduli spaces now depends on the line bundle L and, for the case where L is the canonical bundle ω_X, these pairs are Higgs bundles and we obtain:

Theorem 4.8 ([72]) *The space $\mathcal{M}(r, d, \omega_X)$ is a quasi-projective variety of complex dimension $\dim_{\mathbb{C}} \mathcal{M}(r, d, \omega_X) = (r^2 - 1)(2g - 2)$.*

As we saw in Sect. 4.3.2, we can obtain the moduli space of Higgs bundles from the representations point of view. Let X be a compact Riemann surface and let Γ be a universal central extension of the fundamental group $\pi_1(X, x_0)$. Consider the space of representations $\text{Hom}(\Gamma, GL(r, \mathbb{C}))$ up to the equivalence relation given by conjugation in $GL(r, \mathbb{C})$. It happens now that, in order to have a Hausdorff quotient, we need to restrict to certain representations which are *reductive*, i.e., direct sums of irreducible representations. Playing the same correspondence as in Sect. 4.3.2, a notion of reductive connection can be defined as one where each invariant subbundle admits an invariant complement. Donaldson [19] in rank 2 and Corlette [17] in higher rank show that reductive connections with constant central curvature admit harmonic metrics on the underlying smooth bundle $\mathbb{E}$, which decompose the reductive connection into a unitary connection (yielding holonomy in $U(r)$ and therefore defining a holomorphic vector bundle) plus a 1-form related to the Higgs field. This is how representations into $GL(r, \mathbb{C})$ are related to Higgs bundles.

We define the *moduli space of reductive representations* of degree d of a universal central extension Γ of the fundamental group of a compact Riemann surface X into the general linear group $GL(r, \mathbb{C})$ as the GIT quotient

$$\mathcal{R}^H(r, d) = \mathrm{Hom}_{red}^d (\Gamma, GL(r, \mathbb{C})) \mathbin{/\mkern-6mu/} GL(r, \mathbb{C}) ,$$

and the analogous *moduli space of irreducible representations*:

$$\mathcal{R}_{irr}^H(r, d) = \mathrm{Hom}_{irr}^d (\Gamma, GL(r, \mathbb{C})) / GL(r, \mathbb{C}) .$$

There exists a homeomorphism with the moduli space of Higgs bundles:

$$\mathcal{R}^H(r, d) \simeq \mathcal{M}^H(r, d)$$

restricting to a homeomorphism between irreducible representations and stable Higgs bundles:

$$\mathcal{R}_{irr}^H(r, d) \simeq \mathcal{M}_s^H(r, d) .$$

This correspondence is usually referred as the *non-abelian Hodge correspondence* because it extends Hodge theory in the abelian cohomology groups $H^1(X, \mathbb{C})$ to the non-abelian case $H^1(X, GL(r, \mathbb{C}))$.

Higgs bundles can be generalized to higher dimensional varieties X. Given X a smooth complex projective variety of dimension n, a *Higgs bundle* is a pair (E, φ) where E is a locally free sheaf over X and a *Higgs field* is a morphism $\varphi : E \to E \otimes \Omega_X^1$ verifying $\varphi \wedge \varphi = 0$. These pairs (E, φ) can be thought as coherent sheaves on the cotangent bundle $TX^\vee$. This definition is due to Simpson, who constructs a moduli space for semistable Higgs bundles with this point of view in [86].

The total space of $TX^\vee$ can be thought as having two kinds of coordinates, those coming from X and those coming from the fibers. Functions on the coordinates of X give the holomorphic structure of E, and functions on the fiber encode the multiplication prescribed by the Higgs field.

Chapter 5
Unstability Correspondence

In this chapter we survey results on different correspondences of the stability notion and the GIT picture, at the level of maximal unstability provided by the Harder-Narasimhan filtration.

We begin in Sect. 5.1 by recalling the moduli space of holomorphic vector bundles from Sect. 4.1. Starting with a stability notion for vector bundles in terms of their slope, a Harder-Narasimhan filtration exists as in Sect. 4.2. From the GIT picture, a result by Kempf [56] assures that there exists a maximal 1-parameter subgroup optimizing the Hilbert-Mumford criterion characterizing GIT stability, and this 1-parameter subgroup defines a filtration of the object. Finally we show a correspondence between these two notions. After this, correspondences for other bundle-related moduli problems are explained, say holomorphic pairs, Higgs bundles, and rank 2 tensors.

In Sect. 5.2 a similar correspondence is performed in a different setting, the moduli space of quiver representations on vector spaces. The same Kempf result applies but with other particularities. Finally, in Sect. 5.3 we study a generalization of quiver representations, (G, h)-constellations, for which a stability condition with infinitely many parameters is needed and the previous results hold just asymptotically: instead of an actual correspondence we get an asymptotic convergence.

5.1 Correspondence for Vector Bundles

Recall from Chap. 4 the GIT constructions of moduli spaces classifying some type of algebro-geometric objects modulo an equivalence relation. Usually, we have to impose stability conditions on the objects we classify in order to obtain a space with appropriate properties where each point corresponds to an equivalence class of objects. By rigidifying the objects, which typically involves adding a piece of data to the object we want to parameterize (in the example of the moduli space

© The Author(s), under exclusive license to Springer Nature Switzerland AG 2021
A. Zamora Saiz, R. A. Zúñiga-Rojas, *Geometric Invariant Theory, Holomorphic Vector Bundles and the Harder-Narasimhan Filtration*, SpringerBriefs in Mathematics, https://doi.org/10.1007/978-3-030-67829-6_5

of bundles this piece of data is the isomorphism between V and $H^0(X, E(m)))$, we realize them as points in a finite dimensional parameter space. The freedom in the choice of the additional data corresponds to the action of a group. Mumford's GIT [66] enables then to undertake a quotient, obtaining a projective variety which is a moduli space classifying the objects in the moduli problem. In every moduli problem using GIT, at some point one has to prove that both notions of stability do coincide, then the semistable objects correspond to GIT semistable points and the unstable ones are related to the GIT unstable ones. This eventual identification, in the moduli of vector bundles, happens for a large value of the twist m.

By the Hilbert-Mumford criterion (c.f. Theorem 3.3) we can characterize GIT stability through a numerical function on 1-parameter subgroups, which turns out to be positive or negative when the 1-parameter subgroup destabilizes a point or not, in the sense of GIT. Besides, when a point is GIT unstable, we are able to talk about *degrees of unstability* corresponding to 1-parameter subgroups which are more destabilizing than others (recall Sect. 3.5). Based on the work of Mumford [66], Tits [88], and Kempf [56], among others, we can measure this by means of a rational function on the space of 1-parameter subgroups, whose numerator is the numerical function of the Hilbert-Mumford criterion and the denominator is a norm in the space of 1-parameter subgroups. We choose this norm to avoid rescaling the numerical function. By a theorem of Kempf [56] there exists a unique 1-parameter subgroup giving a maximum for this function and representing the maximal GIT unstability, an idea which can also be seen from the differential and symplectic points of view as the direction of maximal descense (recall Sect. 3.5 and see references [23, 57, 59]). Hence, an unstable object gives a GIT unstable point for which there exists a unique 1-parameter subgroup GIT maximal destabilizing. From this 1-parameter subgroup it is possible to construct a filtration by subobjects of the original unstable object, which makes sense to ask whether it coincides with the Harder-Narasimhan filtration in cases where it is already known, or if it is able to provide a new notion of such filtration in other cases. In the following sections, we will see how to implement this correspondence program for different moduli problems, and stress their similarities and differences.

A treatment of the same idea of optimal destabilizing vectors relating them with Harder-Narasimhan filtrations, but from the gauge theoretical point of view, can be found in [15, 16]. Other perspective of this correspondence appears in [49] by producing moduli spaces of unstable loci with fixed Harder-Narasimhan filtration in the GIT problem. Extensions of this idea to quiver representations and close relationships with symplectic quotients can also be seen in [47, 48]. The theory of destabilizing objects in moduli theory [35, 36] is a deep technical generalization to the language of stacks, having the discussion of Sect. 5.1.1 as a particular case.

5.1.1 Main Correspondence: Holomorphic Vector Bundles

In [28] the main case of this correspondence is discussed for torsion-free coherent sheaves over projective varieties, which is the generalization of the moduli space of holomorphic vector bundles over smooth complex projective curves to higher dimensional varieties. The construction is due to Gieseker [24] for algebraic surfaces and extended to the higher dimensional case by Maruyama [61]. The main difference with the moduli of vector bundles over curves is that the stability condition is expressed by means of a polynomial, the Hilbert polynomial $P_E(m) = \chi(E(m))$, encoding the information of all Chern classes, not just the degree (see Sect. 2.3.3). A torsion-free coherent sheaf is said to be semistable if for every nonzero proper subsheaf $F \subsetneq E$ the following holds for large values of m:

$$\frac{P_F(m)}{\mathrm{rk}(F)} \leq \frac{P_E(m)}{\mathrm{rk}(E)} \; ,$$

and stable if the inequality is strict for all nonzero proper subsheaves. For the sake of simplicity, we will state the results for vector bundles over smooth complex projective curves, referring the reader to [28] for the higher dimensional case.

Let X be a smooth complex projective curve embedded in a projective space by means of a very ample line bundle $O_X(1)$. Fix the rank r and the degree d, and consider the moduli problem of classifying holomorphic vector bundles E over X with rank r and degree d, and fixed determinant $\det(E) = L$, whose solution is given by the GIT construction of the moduli space in Sect. 4.1. Recall the notion of stability (c.f. Definition 4.1) and the existence of a Harder-Narasimhan filtration for these objects (c.f. Theorem 4.1).

We want to establish a relationship between an unstable bundle and a GIT-unstable point in the GIT quotient such that the Harder-Narasimhan filtration can be recovered from the GIT picture, somehow. To this purpose, let us start with an unstable holomorphic bundle E over X. When it is shown in the GIT construction of Sect. 4.1 that we choose an integer m, sufficiently large, for the family of all semistable bundles to be bounded, it is clear that some unstable bundles will show up in the Quot-scheme and need to be removed to recover a moduli space of just semistable bundles. Then, consider an integer m_1 given by the maximum of this m, and the m_0 in Theorem 2.2 such that our unstable E is also m_0-regular; hence E, together with an isomorphism α between a complex vector space V and the space of global sections $H^0(X, E(m_1))$, will be a point $[q']$ contained in the Quot-scheme Z.

The moduli space is the GIT quotient by the action of the group $\mathrm{SL}(n, \mathbb{C})$, and during the GIT process, 1-parameter subgroups $\rho : \mathbb{C}^* \to \mathrm{SL}(n, \mathbb{C})$ distinguish between GIT stable and GIT unstable points. This is given by the calculation of the minimum weight in the Hilbert-Mumford criterion (c.f. Theorem 3.3). For a GIT unstable point $[q']$, it is stated that there is at least one 1-parameter subgroup giving a positive weight $\vartheta([q'], \rho)$. Then, it makes sense to ask whether there is one distinguished 1-parameter subgroup maximizing this weight which, in turn, would achieve the idea of being maximal destabilizing. Before doing this, we need to calibrate the weights to avoid rescaling and therefore obtain a well-defined maximum.

Let $\mathcal{R}(G)$ be the set of all 1-parameter subgroups of G. Define a *length* on $\mathcal{R}(G)$ to be the norm associated to a positive definite integral-valued bilinear form $\langle\ ,\ \rangle$ on $\mathcal{R}(T)$ invariant by the Weyl group, where T is a maximal torus of G, such that

$$\langle\rho, \rho\rangle = \|\rho\|^2 .$$

If G is simple, as it is $\mathrm{SL}(n, \mathbb{C})$, all lengths will be given by multiples of the Killing form (see Sect. 2.1.2), thus the length is unique up to a scalar. Then, for the case $G = \mathrm{SL}(n, \mathbb{C})$ the 1-parameter subgroups ρ can be diagonalized with diagonal $(t^{\rho_1}, t^{\rho_2}, \ldots, t^{\rho_n})$ and the only choice of length is, up to scalar (c.f. Example 3.9):

$$\|\rho\| = \sqrt{\langle\rho, \rho\rangle} = \sqrt{\frac{1}{2}\,\mathrm{tr}\left(\mathrm{ad}\left(\rho_*\frac{d}{dt}\right)^2\right)} = \sqrt{\sum_{i=1}^{n}\rho_i^2} .$$

Then, let us define the *Kempf function* in a GIT problem:

$$\Phi_x(\rho) = \frac{\vartheta(x, \rho)}{\|\rho\|} ,$$

where the numerator is the weight of the action of ρ over x, as in the Hilbert-Mumford criterion (c.f. Theorem 3.3), and the denominator is the length as defined before. The numerator of the Kempf function is precisely the speed of unstability, in the sense that the higher the exponents (thus a greater weight $\vartheta(x, \rho)$) the faster the limit of the 1-parameter subgroup goes toward zero in the affine cone (see Theorem 3.2 and Sect. 3.2), and the denominator serves for normalizing the weight. Note that, if we multiply the exponents in a 1-parameter subgroup of $\mathrm{SL}(n, \mathbb{C})$ by a quantity a, the weight $\vartheta(x, \rho)$ gets multiplied by this a but also the norm in the denominator, therefore the Kempf function gets rid of this rescaling giving a well-defined maximum. Note that the Kempf function coincides with the function in Sect. 3.5 coming from [23].

Finally we can state Kempf's result on maximal destabilizing 1-parameter subgroups.

Theorem 5.1 ([56, Theorem 2.2]) *Let x be a GIT unstable point. There exists a unique 1-parameter subgroup ρ, up to conjugation by an element of the parabolic subgroup defined by ρ, such that the Kempf function $\Phi_x(\rho) = \dfrac{\vartheta(x, \rho)}{\|\rho\|}$ achieves its maximum.*

Let $[q']$ be the GIT point associated to the unstable (E, α), and let ρ be the maximal destabilizing 1-parameter subgroup given by Theorem 5.1. This 1-parameter subgroup ρ produces a weighted flag of vector subspaces (as we saw in Sect. 4.1):

$$0 \subset V_1^{m_1} \subset V_2^{m_1} \subset \cdots \subset V_t^{m_1} \subset V_{t+1}^{m_1} = V$$

which, by the isomorphism $V \simeq H^0(X, E(m_1))$, is a filtration of vector subspaces of global sections and can be evaluated to produce a filtration of subsheaves of E:

$$0 \subset E_1^{m_1} \subset E_2^{m_1} \subset \cdots \subset E_t^{m_1} \subset E_{t+1}^{m_1} = E \, .$$

We call this last filtration the m_1-*Kempf filtration* of the unstable bundle E. The integer m_1 resembles on the twist of the GIT problem for the Quot-scheme to contain all semistable bundles plus our unstable one, and recall that because of the regularity definition (c.f. Sect. 2.3.2) this works for all integers $\ell \geq m_1$. Then we could have chosen a different integer $\ell > m_1$ and obtain another ℓ-Kempf filtration.

It is natural to think that this Kempf filtration, coming from the GIT picture in a maximal destabilizing way, should be related to the Harder-Narasimhan filtration in Theorem 4.1. To show this, on the one hand, the whole process so far depends on an integer m related to the embedding of the Quot-scheme in a projective space as we have said. And for different integers m, in principle, different m-Kempf filtrations arise. On the other hand, we need to show that this filtration satisfies the properties of the Harder-Narasimhan filtration, which makes it unique among all possible filtrations of E.

Theorem 5.2 ([28, Proposition 5.3 and Corollary 6.4]) *Let E be an unstable holomorphic vector bundle. Given an integer m sufficiently large, the Kempf filtration obtained by evaluating the filtration of vector subspaces of $V \simeq H^0(X, E(m))$ does not depend on m. Moreover, this filtration satisfies the properties of the Harder-Narasimhan filtration of E, therefore it coincides with it.*

The proof of this result is rather technical and we include here a sketch of the steps toward it, for completeness.

The Kempf function can be seen as a function over a convex set (c.f. [28, Theorem 2.2 and Proposition 3.2]), hence it attains a maximum at some point, which satisfies certain convexity properties. With the translation into the vector bundles framework, several statements lead to show that the m-Kempf filtration, for an m sufficiently large, does not depend on m. The argument in all the steps is usually the following: the Kempf filtration achieves the maximum of the Kempf function over all possible filtrations; besides, this maximal filtration needs to correspond to the optimum of the function over the convex set, an optimal point satisfying two convexity properties; therefore if the Kempf filtration does not satisfy these properties, it cannot be the maximal filtration.

Le Potier-Simpson estimates on bounds for the dimension of the space of global sections (c.f. [60], [86, Corollary 1.7] and [53, Lemma 2.2]) allow to prove that we can chose another integer $m_2 \geq m_1$ such that all subbundles appearing on the m-Kempf filtrations, for $m \geq m_2$, have their numerical invariants (degree and thus slope) bounded by below (see [28, Propositions 3.8 and 3.9]). Then, it is satisfied a regularity property for the family of all subbundles appearing in the different m-

Kempf filtrations, for $m \geq m_2$: all E_i^m are generated by global sections and they lose their higher cohomology (see Sect. 2.3.2).

After this, it can be proved (c.f. [28, Proposition 3.10]) that, actually, not only the subspaces $V_i \subset H^0(X, E_i^m)$ generate the subbundles $E_i^m \subset E$. Moreover, there exists another integer $m_3 \geq m_2$ such that, for $m \geq m_3$, we have $V_i = H^0(X, E_i^m)$ indeed. From this, the m-regularity property allows to fix the vector space V once and for all, and not to vary with m anymore. Then, the different filtrations of V are completely related to the cohomological invariants of the subbundles, and the finiteness of these invariants help to prove that the different m-Kempf filtrations can be reduced to a unique possible maximal filtration (c.f. [28, Propositions 4.1 and 4.3]), the *Kempf filtration*, keeping no more track of the integer m.

Finally, it happens that the two convexity properties of the optimal vector of the Kempf function over the convex set are closely related to those of the Harder-Narasimhan filtration. These are that the graph associated to the filtration is convex (c.f. [28, Lemma 3.4]) and that this graph is also the convex envelope of every refinement (c.f. [28, Lemma 3.5]). These two properties are precisely that the slopes of the filtration of vector bundles are decreasing and that the quotients are semistable (if not, a destabilizing subbundle of the quotient gives rise to a point above the graph, breaking its convexity). The graph is then associated to the Harder-Narasimhan polygon (see Sect. 4.2) and the properties are the ones defining the Harder-Narasimhan filtration in Theorem 4.1. Therefore the Kempf filtration, which is the filtration giving a maximal way of destabilizing a GIT unstable point through the Kempf's 1-parameter subgroup, and now unique independently of the integer m and the vector space V of the GIT quotient, coincides with the Harder-Narasimhan filtration, completing the proof of Theorem 5.2.

5.1.2   Other Correspondences for Augmented Bundles

The previous construction can be carried out for other moduli problems of augmented bundles which are vector bundles together with bundle-related morphisms. A general treatment of GIT problems with augmented bundles can be read in [79].

We call a *holomorphic pair* to the pair

$$(E, \phi : E \longrightarrow O_X)$$

consisting on a rank r vector bundle E with fixed determinant $\det(E) \cong L$ over a smooth complex projective variety X, and a morphism ϕ to the trivial line bundle O_X. These objects are very related to Bradlow pairs consisting of a vector bundle together with a global section $s \in H^0(X, E)$ (see [12]).

There is a notion of stability for holomorphic pairs depending on a parameter. When X is a smooth complex projective curve, this parameter is a rational number δ but, in the higher dimensional case with $\dim X = n$ and Gieseker stability given by

Hilbert polynomials, the parameter takes the form of a $(n-1)$-degree polynomial $\delta(m)$. For $n = 1$, a holomorphic pair (E, ϕ) is δ-*semistable* if, for all nonzero subpairs $(F, \phi|_F) \subset (E, \phi)$ it is

$$\frac{\deg(F) + \delta\epsilon(F)}{\mathrm{rk}(F)} \leq \frac{\deg(E) + \delta}{\mathrm{rk}(E)} \; ,$$

where $\epsilon(F) = 1$ if $\phi(F) \neq 0$ generically, and $\epsilon(F) = 0$ otherwise. With these ingredients, a moduli space for semistable holomorphic pairs is constructed, by using Geometric Invariant Theory (see [51, 52] and also [26, 78] for generalizations).

In the GIT construction, again, all δ-semistable pairs satisfy an m-regularity property such that they can be embedded in certain Quot-scheme acted by $\mathrm{SL}(n, \mathbb{C})$, and we need to take the quotient by this group to obtain the moduli space. Starting from a δ-unstable holomorphic pair $(E, \phi : E \to \mathcal{O}_X)$ and an integer m such that this δ-unstable pair lays in the Quot-scheme, Kempf's theorem applies to show that there is a unique 1-parameter subgroup ρ maximal destabilizing in the sense of GIT, that is, achieving the maximum for the Kempf function in this case. This ρ provides a filtration of V:

$$0 \subset V_1^m \subset V_2^m \subset \cdots \subset V_t^m \subset V_{t+1}^m = V$$

and, by evaluating, a filtration of holomorphic pairs, the m-Kempf filtration of (E, ϕ):

$$0 \subset (E_1^m, \phi|_{E_1^m}) \subset (E_2^m, \phi|_{E_2^m}) \subset \cdots \subset (E_t^m, \phi|_{E_t^m}) \subset (E_{t+1}^m, \phi|_{E_{t+1}^m}) = (E, \phi) \; .$$

Following the same program, it can be shown that there exists an integer m sufficiently large such that the m-Kempf filtration is independent on m (c.f. [29, Theorem 3.2]), and that this filtration coincides with the Harder-Narasimhan filtration (c.f. [29, Theorem 3.3]) which is shown to exist, to be unique, and to have the expected analogous properties (c.f. [29, Theorem 1.4]).

Another interesting case happens with Higgs bundles from Sect. 4.4. Following Simpson's construction [86] a Higgs bundle (E, φ) over X is seen as a coherent sheaf on the cotangent bundle of X:

$$\pi : TX^\vee \longrightarrow X \; ,$$

supported on the spectral curve which is the curve of eigenvalues of the Higgs field φ. As we pointed out at the end of Sect. 4.4.2, to define the rings of transition functions on the total space of $TX^\vee$ we need to specify how to multiply a section by the vertical variables, which is given by the Higgs field, by definition.

In this case, the same correspondence is performed for the moduli space of Higgs bundles (see [95, Section 2.3]), which is indeed classifying pure (i.e., supported in a lower-dimensional variety) coherent sheaves of fixed numerical invariants. A filtration of pure subsheaves

$$0 \subset \mathcal{E}_1 \subset \mathcal{E}_2 \subset \cdots \subset \mathcal{E}_t \subset \mathcal{E}_{t+1} = \mathcal{E}$$

is found by Kempf theorem and proved to be independent of the integer in the GIT moduli construction, eventually. This filtration is push-forwarded:

$$0 \subset \pi_*(\mathcal{E}_1) \subset \pi_*(\mathcal{E}_2) \subset \cdots \subset \pi_*(\mathcal{E}_t) \subset \pi_*(\mathcal{E}_{t+1}) = \pi_*(\mathcal{E}) = (E, \varphi)$$

to obtain a filtration of Higgs subbundles:

$$0 \subset (E_1^m, \varphi|_{E_1^m}) \subset (E_2^m, \varphi|_{E_2^m}) \subset \cdots \subset (E_t^m, \varphi|_{E_t^m}) \subset (E_{t+1}^m, \varphi|_{E_{t+1}^m}) = (E, \varphi) ,$$

which coincides with the Harder-Narasimhan filtration of (E, φ) (see [5]).

There are moduli problems where there is no notion of what a Harder-Narasimhan filtration should be a priori. In these cases, the kind of correspondences that we present can help to define a Harder-Narasimhan filtration as that filtration coming from the maximal destabilizing 1-parameter subgroup in the GIT quotient. We call a *rank* 2 *tensor* the pair consisting of

$$(E, \ \tau : E \otimes \cdots \otimes E \longrightarrow M) ,$$

where E is a rank 2 coherent torsion-free sheaf over a smooth complex projective variety X, M is a line bundle over X, and τ is a symmetric morphism $\tau : E^{\otimes s} \to M$ (i.e., fiberwise, it is a symmetric multilinear map). In this case, the filtrations for these objects consist only on line subbundles L and the corresponding restricted morphisms:

$$0 \subset (L, \tau|_L) \subset (E, \tau) .$$

When X is a curve, the projectivization of the total space of a rank 2 bundle gives a ruled algebraic surface $\mathbb{P}(E)$. On the other hand, a symmetric morphism τ is, fiberwise, a symmetric multilinear map

$$\tau_x : (\mathbb{C}^2)^{\otimes s} \longrightarrow \mathbb{C}$$

which can be represented by a degree s homogeneous polynomial $\tau_x(X, Y) = \sum_{i=0}^{s} a_i(x) X^i Y^{s-i}$ vanishing at s points on $\mathbb{C}^2 \simeq \mathbb{P}_{\mathbb{C}}^1$, counted with multiplicity. Then, when moving x along the curve X, it defines an s-degree covering $\tilde{X} \to X$ lying on the ruled surface $\mathbb{P}(E)$ (Fig. 5.1).

A moduli space for δ-semistable tensors and G-principal bundles is constructed in [26, 27], where δ is a parameter (or a polynomial for higher dimensional X) similar to the one for holomorphic pairs. In [97], it is defined a notion of stable covering as the covering $\tilde{X} \to X$ associated to a δ-stable symmetric rank 2 tensor (E, τ), and the previous treatment leads to find a subtensor $(L, \tau|_L) \subset (E, \tau)$,

Fig. 5.1 Rank 2 tensor (E, τ). The vertical lines of $\mathbb{P}(E)$ are the rank 2 fibers of E over X and define a ruled surface. The morphism τ defines a (degree $s = 4$) covering $\tilde{X} \to X$. The line subbundle $L \subset E$ defines a section lying on the ruled surface. At the point $x \in X$, τ_x has four zeros two of them simple and one double, and this double zero is crossed by L at the fiber E_x

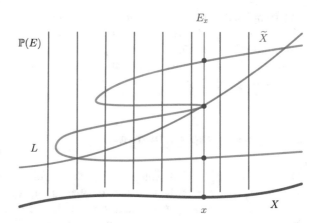

maximal destabilizing from the GIT point of view. For a sufficiently large m-twist of the GIT problem this subtensor does not depend on the integer of the embedding and defines a 1-step filtration, which we declare to be the *Harder-Narasimhan filtration* of (E, τ). This way we find a main application of the techniques developed in a problem where a Harder-Narasimhan filtration is not known a priori.

Besides this, we can characterize geometrically the maximal destabilizing subbundle $L \subset E$. The minimal weight in the Hilbert-Mumford criterion of Theorem 3.3 in this problem is of the form (see [97])

$$\vartheta(x, \rho) = A + \delta B \ ,$$

where A and B are numbers. For A we have

$$A = 2 \deg(L) - \deg(E) = \mathrm{rk}(E) \cdot (\mu(L) - \mu(E)) \ ,$$

then maximizing A is equivalent to find a destabilizing line subbundle $L \subset E$ with the greatest possible slope $\mu(L) > \mu(E)$. Given that a line bundle defines a section in the ruled surface $\mathbb{P}(E)$, this number A can be characterized in terms of intersection theory for algebraic surfaces. For B, it turns out that it will be $B > 0$ when, generically, the point defined by L in the fiber of the ruled surface coincides with one of the s zeros of τ, and that zero has multiplicity greater than $s/2$. This way, the term B helps in favor of the unstability if the configuration of points defined by the morphism τ is generically unstable (in terms of the Example 3.5) and that movable zero along X, of multiplicity higher than $s/2$, can be realized as the divisor of a line subbundle. This brings together two seminal treatments of the notion of stability by Hilbert and Mumford [44, 66], in this particular example.

Coming from these ideas it could be thought that a similar construction can be performed for principal G-bundles (see [73]). The notion of Harder-Narasimhan filtration for principal bundles is achieved by means of the canonical reduction, a reduction to a parabolic subgroup $P \subset G$. In [9], examples of orthogonal and symplectic bundles over surfaces are shown where the Harder-Narasimhan filtration of its underlying vector bundle does not correspond to any parabolic reduction of the orthogonal or symplectic bundle. This result prevents us from trying to relate unstability of G-bundles with unstability of the underlying bundle so straightforwardly, as it happens with stability.

5.2   Quiver Representations

There are other different categories, such as quiver representations, where the ideas from Sect. 5.1 apply. Consider representations of a finite quiver Q on finite dimensional k-vector spaces, where k is an algebraically closed field of arbitrary characteristic. We recall the construction of a moduli space for these objects by King [58] and associate to an unstable representation an unstable point, in the sense of GIT, in a parameter space where a group acts. Then, the 1-parameter subgroup of Kempf gives a filtration of subrepresentations and we prove that it coincides with the Harder-Narasimhan filtration for that representation (see [97]).

Let Q be a finite *quiver* given by a finite set of vertices $Q_0 = \{v_i\}_{i \in I}$ and a finite set of arrows $Q_1 = \{\alpha : v_i \to v_j\}$. Denote by $\mathbb{Z}Q_0$ the free abelian group generated by Q_0. Let mod kQ be the category of finite dimensional *representations* of Q over k, an algebraically closed field of arbitrary characteristic. The objects of this category are tuples

$$M = \left((M_v)_{v \in Q_0}, (M_\alpha : M_{v_i} \longrightarrow M_{v_j})_{\alpha \in Q_1}\right)$$

of finite dimensional vector spaces and k-linear maps. We represent the numerical invariants of a representation by means of their *dimension vector*:

$$\underline{\dim} M = \sum_{v \in Q_0} \dim_k M_v \cdot v \in \mathbb{N}Q_0.$$

Given rational numbers θ_v for each $v \in Q_0$, define the linear function

$$\theta : \mathbb{Z}Q_0 \longrightarrow \mathbb{Q}, \quad \theta(M) = \theta(\underline{\dim} M) = \sum_{v \in Q_0} \theta_v \dim_k M_v.$$

Also, consider a set of strictly positive numbers σ_v for each $v \in Q_0$, and define the (strictly positive) linear function

$$\sigma : \mathbb{Z}Q_0 \longrightarrow \mathbb{Q}_{>0}, \quad \sigma(M) = \sigma(\underline{\dim} M) = \sum_{v \in Q_0} \sigma_v \dim_k M_v,$$

called *the total dimension of* M. With these two functions, define the (θ, σ)-slope of a representation M by

$$\mu_{(\theta,\sigma)}(M) := \frac{\theta(M)}{\sigma(M)} .$$

Definition 5.1 A representation M is (θ, σ)-semistable if for every nonzero proper subrepresentation $M' \subsetneq M$ we have

$$\mu_{(\theta,\sigma)}(M') \le \mu_{(\theta,\sigma)}(M).$$

If the inequality is strict for every nonzero proper subrepresentation, we say that M is (θ, σ)-stable.

Definition 5.1 rephrases King's definition (c.f. [58, Definition 1.1]) by adding the total dimension and the slope condition, instead of just the numerator. This allows to obtain, besides a moduli space for these objects, the existence and uniqueness of a Harder-Narasimhan filtration, which is automatic by using [76] because quiver representations are an abelian category where the slope plays an additive role in exact sequences.

Theorem 5.3 ([76, Theorem 2], [74, Lemma 4.7]) *Let θ and σ be two linear functions (σ being strictly positive). Every representation M of a finite quiver Q over k has a unique* Harder-Narasimhan filtration

$$0 = M_0 \subsetneq M_1 \subsetneq M_2 \subsetneq \cdots \subsetneq M_t \subsetneq M_{t+1} = M$$

verifying the following properties, where $M^i := M_i/M_{i-1}$:

1. $\mu_{(\theta,\sigma)}(M^1) > \mu_{(\theta,\sigma)}(M^2) > \cdots > \mu_{(\theta,\sigma)}(M^t) > \mu_{(\theta,\sigma)}(M^{t+1})$.
2. *The quotients M^i are (θ, σ)-semistable.*

To construct a moduli space of quiver representations (c.f. [59]) let us fix the numerical invariants, that is fix a dimension vector $d \in \mathbb{Z}Q_0$ and fix k-vector spaces M_v of dimension d_v for all $v \in Q_0$. And let us set a stability condition given by linear functions θ and σ, with σ strictly positive.

Representations are parametrized by the affine k-vector space

$$\mathcal{R}_d(Q) := \bigoplus_{\alpha:v_i \to v_j \in Q_1} \mathrm{Hom}_k(M_{v_i}, M_{v_j}) ,$$

where the reductive linear algebraic group $G_d := \prod_{v \in Q_0} \mathrm{GL}(d_v, k)$ acts by

$$(g_v)_{v \in Q_0} \cdot (M_\alpha)_{\alpha \in Q_1} = \left(g_{v_j} M_\alpha g_{v_i}^{-1} \right)_{\alpha:v_i \to v_j \in Q_1} .$$

Then, the orbits of M by the action of G_d in the space $\mathcal{R}_d(Q)$ correspond bijectively to the isomorphism classes $[M]$ of k-representations of Q with dimension vector d.

Next, we would like to use Geometric Invariant Theory to take the quotient of $\mathcal{R}_d(Q)$ by G_d. However, since the parameter space is affine and not projective, there are not enough invariant functions and we will consider semi-invariants as we did in Example 3.2. We lift the action of G_d on $\mathcal{R}_d(Q)$ by a character χ to the trivial line bundle $\mathcal{R}_d(Q) \times k$. In this case, the action of G_d has a kernel, i.e. the subgroup of diagonal scalar matrices $\Delta = \{(\alpha\, I, \ldots, \alpha\, I) : \alpha \in k^*\} \subset G_d$ acts trivially on $\mathcal{R}_d(Q)$. Then we need to choose χ such that Δ acts trivially on the fiber, that is, $\chi(\Delta) = 1$ (see [59] for the modifications caused by this with respect to Example 3.2). This character satisfies the property:

$$\chi_{(\theta,\sigma)}\left((g_v)_{v \in Q_0}\right) := \prod_{v \in Q_0} \det\,(g_v)^{(\theta(d)\sigma_v - \sigma(d)\theta_v)} \,.$$

Then, recall that a point $x \in \mathcal{R}_d(Q)$ is $\chi_{(\theta,\sigma)}$-semistable if there exists a G_d-semi-invariant f with respect to the character $\chi_{(\theta,\sigma)}^m$, with $m \geq 1$, such that $f(x) \neq 0$. If we denote by $\mathcal{R}_d^{ss}(Q)_{\chi_{(\theta,\sigma)}}$ the subset of $\chi_{(\theta,\sigma)}$-semistable points, the GIT quotient is given by

$$\mathcal{R}_d^{ss}(Q)_{\chi_{(\theta,\sigma)}} /\!\!/ G_d = \mathrm{Proj}\left(\bigoplus_{m \geq 0} k\,[\mathcal{R}_d(Q)]^{G_d}_{\chi_{(\theta,\sigma)}^m}\right),$$

where $k[\mathcal{R}_d(Q)]^{G_d}_{\chi_{(\theta,\sigma)}^m}$ is the set of G_d-semi-invariants for each m. This result guarantees the correspondence between (θ, σ)-semistable representations and GIT semistable points to construct a moduli space for semistable representations.

Proposition 5.1 ([95, Proposition 2.2]) *A point $x_M \in \mathcal{R}_d(Q)$ corresponding to a representation $M \in \mathrm{mod}\, kQ$ is $\chi_{(\theta,\sigma)}$-semistable for the action of G_d if and only if M is (θ, σ)-semistable. The same correspondence holds for $\chi_{(\theta,\sigma)}$-stable and (θ, σ)-stable, respectively.*

Theorem 5.4 ([58, Proposition 4.3], [74, Corollary 3.7]) *The quotient of the set of $\chi_{(\theta,\sigma)}$-semistable points:*

$$\mathcal{M}_d^{(\theta,\sigma)}(Q) = \mathcal{R}_d^{ss}(Q)_{\chi_{(\theta,\sigma)}} /\!\!/ G_d \,,$$

is a moduli space parametrizing S-equivalence classes of (θ, σ)-semistable representations of Q with dimension vector d.

To continue, we need to apply the Hilbert-Mumford criterion to characterize GIT stability. In this moduli problem, a 1-parameter subgroup of $G_d = \prod_{v \in Q_0} \mathrm{GL}(d_v, k)$ is a nontrivial homomorphism $\rho : k^* \to G_d$ which, after taking bases for the vector spaces M_v on each vertex, are of the form:

$$\rho(t) = \begin{pmatrix} t^{\rho_{v_1,1}} & & \\ & \ddots & \\ & & t^{\rho_{v_1,t_1+1}} \end{pmatrix} \times \cdots \times \begin{pmatrix} t^{\rho_{v_s,1}} & & \\ & \ddots & \\ & & t^{\rho_{v_s,t_s+1}} \end{pmatrix}.$$

Defining $\vartheta_{\chi_{(\theta,\sigma)}}(x_M, \rho) = a$, where a is the weight of the action of ρ on a point $x_M \in \mathcal{R}_d(Q)$ (i.e., ρ acts on the fiber of the trivial line bundle over $x_0 := \lim_{t \to 0} \rho(t) \cdot x_M$ as multiplication by t^a), a point $x_M \in \mathcal{R}_d(Q)$ corresponding to a representation M turns out to be $\chi_{(\theta,\sigma)}$-semistable if and only if every 1-parameter subgroup ρ of G_d for which $\lim_{t \to 0} \rho(t) \cdot x_M$ exists satisfies $\vartheta_{\chi_{(\theta,\sigma)}}(x_M, \rho) \leq 0$ (c.f. [58, Proposition 2.5] and Example 3.4).

The action of ρ breaks each vector space M_v into subspaces of different dimensions, an analog to the moduli of vector bundles, and the quantity a can be written in terms of filtrations of subrepresentations of M:

$$0 = M_0 \subset M_1 \subset M_2 \subset \cdots \subset M_t \subset M_{t+1} = M$$

by saying that (c.f. [95, Proposition 2.6]) x_M is $\chi_{(\theta,\sigma)}$-semistable if and only if every $\rho \in G_d$ defining a filtration of M verifies

$$\vartheta_{\chi_{(\theta,\sigma)}}(x_M, \rho) = \sum_{i=1}^{t+1} \rho_i \cdot \left[\theta(M) \cdot \sigma(M^i) - \sigma(M) \cdot \theta(M^i) \right] \leq 0 \,,$$

where recall that $M^i = M_i / M_{i-1}$.

The group G_d is not simple, so choosing a length on G_d is not unique. Nevertheless, all lengths are given by a weighted sum of the Killing length for each simple factor in the product (recall Sects. 2.1.2 and 5.1.1). Then, to perform the correspondence between the GIT picture and the Harder-Narasimhan filtration, define the following *Kempf function*:

$$\Phi_{x_M}(\rho) = \frac{\vartheta_{\chi_{(\theta,\sigma)}}(x_M, \rho)}{\|\rho\|} = \frac{\sum\limits_{i=1}^{t+1} \rho_i \cdot \left[\theta(M) \cdot \sigma(M^i) - \sigma(M) \cdot \theta(M^i) \right]}{\sqrt{\sum\limits_{i=1}^{t+1} \sigma(M^i) \cdot \rho_i^2}}$$

where note that the norm contains the positive factors σ_v for each vertex. Then, not only the numerator of this Kempf function depends on the components of the stability definition but also the denominator. By Theorem 5.1, there exists a unique 1-parameter subgroup ρ achieving the maximum for the Kempf function.

The end of this case is very straightforward. Let M be a (θ, σ)-unstable representation of Q, then corresponding to a $\chi_{(\theta,\sigma)}$-semistable point $x_M \in \mathcal{R}_d(Q)$.

By Kempf's result, there exists a 1-parameter subgroup ρ, then a filtration of subrepresentations of M:

$$0 \subset M_1 \subset M_2 \subset \cdots \subset M_t \subset M_{t+1} = M$$

maximizing the Kempf function, called the *Kempf filtration* of M. By [28, Lemmas 3.4 and 3.5] this Kempf filtration satisfies the two convexity properties as an optimum over a convex set, which translated into the language of filtrations are precisely the two properties of the Harder-Narasimhan filtration of M in Theorem 5.3. Therefore, the Kempf filtration coincides with the Harder-Narasimhan filtration of a quiver representation, and we achieve to recover this notion from the GIT setting.

> Note how this moduli problem avoids the twisting by the integer to guarantee the boundedness of the semistable objects in the bundle-related moduli problems. Therefore, we work with a unique Kempf filtration during the whole process and the proof needs less technical requirements.

A generalization of this result can be seen in [95, Chapter 3] for Q-sheaves, which are representations of a quiver on the category of coherent sheaves. The moduli construction for this problem can be found in [3].

5.3 (G,h)-Constellations

Many of the previous cases in this chapter, where we have been able to carry the unstability correspondence, fall into abelian categories where the existence of Harder-Narasimhan filtrations is straightforward (see [76]). In other cases where the moduli construction is performed in a non-abelian category, existence and uniqueness of a Harder-Narasimhan filtration is a much harder problem.

In [7] it is studied the moduli problem of (G, h)-constellations, which generalize G-invariant Hilbert schemes. The stability condition for these objects can be seen as an infinite dimensional version of the one for quiver representations in Sect. 5.2. In [87] it is shown that the maximal destabilizing filtration coming from the GIT picture does not necessarily stabilize, that is, eventually it does not change with the embedding in certain Quot-scheme (which depends on a subset D of the set of irreducible representations of G) and defines a unique filtration. However, this filtration asymptotically converges to the Harder-Narasimhan filtration, which is proved to exist.

Let G be an infinite complex reductive algebraic group, whose set of irreducible representations

$$\text{Irr}\, G = \left\{ \xi : G \longrightarrow \text{GL}(V_\xi) \right\}$$

is infinite. Let X be an affine scheme of finite type with an action of the group G. This is equivalent to $X = \mathrm{Spec}(R)$, with R a finitely generated $\mathbb{C}$-algebra with an action of G.

We prescribe a *Hilbert function* $h : \mathrm{Irr}\, G \longrightarrow \mathbb{N}$ and let $\mathcal{F}$ be a G-equivariant coherent $\mathcal{O}_X$-module whose *isotypic decomposition* is

$$H^0(X, \mathcal{F}) = \bigoplus_{\xi \in \mathrm{Irr}\, G} \mathcal{F}_\xi \otimes V_\xi \simeq \bigoplus_{\xi \in \mathrm{Irr}\, G} V_\xi^{h(\xi)},$$

where $\mathcal{F}_\xi = \mathrm{Hom}^G\left(V_\xi, H^0(X, \mathcal{F})\right)$. This means that the global sections of $\mathcal{F}$ decompose as a direct sum of the (infinite) irreducible representations of G with prescribed (finite) multiplicities by the Hilbert function.

Roughly speaking, coherent modules are sheaves that, locally, are finitely generated as modules. The notion of G-equivariant is more general but, if $\mathcal{F}$ is a vector bundle (i.e., a locally free sheaf), being G-equivariant means that each element $g \in G$ induces a linear isomorphism fiberwise $g : \mathcal{F}_x \xrightarrow{\simeq} \mathcal{F}_{g \cdot x}$. Moreover, if $\mathcal{F}$ is a line bundle, G-equivariance is the same as a linearization of the action of G to the total space of $\mathcal{F}$.

Definition 5.2 A (G, h)-*constellation* is a G-equivariant coherent $\mathcal{O}_X$-module $\mathcal{F}$ with multiplicities prescribed by the Hilbert function h.

Example of a (G, h)-constellation

As an example of the definition of a (G, h)-constellation, let $G = \mathbb{C}^*$ whose irreducible representations are in correspondence with $\mathbb{Z}$ because, on each 1-dimensional representation, the variable t acts multiplying by t^r, $r \in \mathbb{Z}$. Consider the action of $\mathbb{C}^*$ on the ring of polynomials in two variables, $\mathbb{C}[X, Y]$, given by

$$t \cdot X = tX \quad \text{and} \quad t \cdot Y = t^{-1}Y,$$

such that the weight of a monomial $X^a Y^b$ is $a - b$. So far, we have infinitely many monomials for each weight; for example, for weight 0:

$$\mathrm{Hom}^{\mathbb{C}^*}(V_0, \mathbb{C}[X, Y]) = \langle 1, XY, X^2Y^2, X^3Y^3, X^4Y^4, \cdots \rangle.$$

Then, we define the affine scheme

$$Z := \mathrm{Spec}\left(\frac{\mathbb{C}[X, Y]}{(XY)}\right)$$

dividing by the ideal (XY) and consider the structure sheaf $\mathcal{F} = O_Z$ itself. We have

$$\frac{\mathbb{C}[X, Y]}{(XY)} \simeq \mathbb{C}[X]_{>0} \oplus \mathbb{C} \oplus \mathbb{C}[Y]_{>0} \simeq \bigoplus_{r \in \mathbb{Z}} V_r,$$

therefore O_Z is a $(\mathbb{C}^*, h)$-constellation with constant Hilbert function $h(r) = 1$, for every $r \in \mathbb{Z} \simeq \mathrm{Irr}\,\mathbb{C}^*$.

As we mentioned before, the stability condition resembles on the quiver representations case, where now the role of the vertices is played by the irreducible representations. The main caveat is that we have an infinite group G, then an infinite number of irreducible representations, which introduce some essential technical difficulties.

Choose rational numbers $\theta = (\theta_\xi)_{\xi \in \mathrm{Irr}\,G}$ such that

$$\sum_{\xi \in \mathrm{Irr}\,G} \theta_\xi h(\xi) = 0$$

and such that the negative ones (i.e., those $\theta_\xi < 0$) are a finite quantity and the positive ones are an infinite quantity. Define D_- and D_+ to be the subsets of $\mathrm{Irr}\,G$ associated to the negative and positive numbers θ_ξ, respectively.

Intending to classify all possible (G, h)-constellations is an unfeasible problem. As in the case of vector bundles we need to assure, somehow, the boundedness of the objects to classify and this is why we restrict to constellations *generated in D_-*, meaning that the sheaf is generated by the sections in the negative subset $D_- \subset D$:

$$\bigoplus_{\xi \in D_-} \mathcal{F}_\xi \otimes V_\xi.$$

Define the θ-*slope* of a constellation $\mathcal{F}$ by

$$\mu_\theta(\mathcal{F}) := \frac{-\theta(\mathcal{F})}{r(\mathcal{F})},$$

where the numerator is the stability function given by

$$\theta(\mathcal{F}) := \sum_{\xi \in \mathrm{Irr}\,G} \theta_\xi \dim \mathcal{F}_\xi = \sum_{\xi \in \mathrm{Irr}\,G} \theta_\xi \dim h(\xi),$$

and the denominator, or *rank*, is the dimension of the negative part D_-:

$$r(\mathcal{F}) := \sum_{\xi \in D_-} \dim \mathcal{F}_\xi = \sum_{\xi \in D_-} \dim h(\xi).$$

Getting all of this together, we define the stability condition.

Definition 5.3 A (G, h)-constellation $\mathcal{F}$ is called μ_θ-*semistable* if for every nonzero proper subsheaf $\mathcal{F}' \subsetneq \mathcal{F}$ generated in D_- we have

$$\mu_\theta(\mathcal{F}') \leq \mu_\theta(\mathcal{F}) .$$

If the previous inequality is strict for every nonzero proper subsheaf, we say that $\mathcal{F}$ is μ_θ-*stable*.

After the slope definition, we are ready to state the existence and uniqueness of a Harder-Narasimhan filtration for constellations. This result is far from being trivial because these time constellations generated in D_- do not form an abelian category, thus the general procedure in [76] does not apply.

Theorem 5.5 ([87, Theorem 1.7]) *Every* (G, h)-*constellation* $\mathcal{F}$ *has a unique* μ_θ-Harder-Narasimhan filtration

$$0 = \mathcal{F}_0 \subsetneq \mathcal{F}_1 \subsetneq \mathcal{F}_2 \subsetneq \cdots \subsetneq \mathcal{F}_t \subsetneq \mathcal{F}_{t+1} = \mathcal{F}$$

verifying the following properties, where $\mathcal{F}^i := \mathcal{F}_i/\mathcal{F}_{i-1}$:

1. *Each* $\mathcal{F}_i$ *is a subsheaf generated in* D_- .
2. $\mu_\theta(\mathcal{F}^1) > \mu_\theta(\mathcal{F}^2) > \cdots > \mu_\theta(\mathcal{F}^t) > \mu_\theta(\mathcal{F}^{t+1})$.
3. *The quotients* $\mathcal{F}^i$ *are* μ_θ-*semistable*.

Next, we recall the construction of a moduli space of constellations from [7], which combines most of the ingredients from the vector bundle and the quiver representation cases. Given θ a stability function and $\mathcal{F}$ a (G, h)-constellation generated in D_-, we define isomorphisms

$$\Upsilon_\xi : \mathbb{C}^{h(\xi)} \xrightarrow{\simeq} \mathcal{F}_\xi$$

for each representation $\xi \in D_-$ in the generating part, to get a surjective quotient

$$q : \mathcal{H} := \left(\bigoplus_{\xi \in D_-} \mathbb{C}^{h(\xi)} \otimes V_\xi \right) \otimes O_X \longrightarrow \mathcal{F}$$

and hence a point

$$[q : \mathcal{H} \longrightarrow \mathcal{F}] \in \mathrm{Quot}^G(\mathcal{H}, h) ,$$

which is the G-invariant Quot-scheme proven to be projective in [54].

The action of the group

$$\Gamma = \left\{ (\gamma_\xi)_{\xi \in D_-} \in \prod_{\xi \in D_-} \mathrm{GL}(h(\xi), \mathbb{C}) \ : \ \prod_{\xi \in D_-} \det(\gamma_\xi) = 1 \right\} ,$$

over $\mathrm{Quot}^G(\mathcal{H}, h)$ encodes the changes of basis for the isomorphisms Υ_ξ, where the determinant condition guarantees to get finite stabilizers. Then, we embed the invariant Quot-scheme by means of a finite set of irreducible representations $D \subset \mathrm{Irr}\, G$:

$$\eta : \mathrm{Quot}^G(\mathcal{H}, h) \hookrightarrow \prod_{\xi \in D} \mathbb{P}^{h(\xi)} .$$

Let $\mathcal{O}_{\mathbb{P}^{h(\xi)}}(\kappa_\xi)$ be ample line bundles on each $\mathbb{P}^{h(\xi)}$, with $\kappa_\xi > 0$, and put them together to get the ample line bundle

$$\mathcal{L} = \eta^* \left(\bigotimes_{\xi \in D} \mathcal{O}_{\mathbb{P}^{h(\xi)}}(\kappa_\xi) \right)$$

on the invariant Quot-scheme, where we linearize the action of Γ by means of the character

$$\chi(\gamma) = \prod_{\xi \in D_-} \det(\gamma_\xi)^{\chi_\xi}$$

with a choice of a parameter χ_ξ for each representation ξ in D_-. With these ingredients we are able to talk about GIT-D-(semi)stable points with respect to the linearized action in $\mathcal{L}_\chi$.

Choosing all parameters κ_ξ, χ_ξ, accordingly with θ_ξ, it can be shown (c.f. [7]) that we can find a subset $D \subset \mathrm{Irr}\, G$ big enough such that

$$\mu_\theta\text{-stable} \implies \text{GIT-}D\text{-stable} \implies \text{GIT-}D\text{-semistable} \implies \mu_\theta\text{-semistable}$$

and the matching between GIT-stability and μ_θ stability is no longer perfect as in the previous moduli problems, finding counterexamples in [87] showing that the previous relations in the diagram are not equivalences.

The GIT stability condition can be turned into a slope condition, namely μ_D-stability, resembling the D-embedding of the invariant Quot-scheme. In a similar way to before, for every (G, h)-constellation $\mathcal{F}$ a unique μ_D-Harder-Narasimhan filtration exists (c.f. [87, Theorem 2.7]):

$$0 \subset \mathcal{G}_1^D \subset \mathcal{G}_2^D \subset \cdots \subset \mathcal{G}_p^D \subset \mathcal{G}_{p+1}^D = \mathcal{F}$$

with analogous properties to the μ_θ-filtration in Theorem 5.5.

In the same vein as with Harder-Narasimhan polygons defined in Sect. 4.2, given the μ_θ-Harder-Narasimhan filtration $\mathcal{F}_\bullet \subset \mathcal{F}$ we associate a θ-*polygon* determined by the points in the plane with coordinates

$$\left(r(\mathcal{F}_i), w_i^\theta\right) , \quad \text{where } w_i^\theta = r(\mathcal{F}_i) \cdot \mu_\theta(\mathcal{F}_i) .$$

And to the μ_D-Harder-Narasimhan filtration $\mathcal{G}_\bullet \subset \mathcal{F}$ we associate the *D-polygon* given by the points in the plane with coordinates

$$\left(r(\mathcal{G}_i), w_i^D\right) , \quad \text{where } w_i^D = r(\mathcal{G}_i) \cdot \mu_D(\mathcal{G}_i) .$$

Finally, we are able to establish an asymptotic relationship between the GIT picture and the stability condition for constellations.

Theorem 5.6 ([87, Theorems 3.3 and 3.7]) *Given a (G, h)-constellation $\mathcal{F}$ and a stability condition θ for which $\mathcal{F}$ is generated in D_-, there exists a finite subset $D \subset \text{Irr } G$ sufficiently large to provide an embedding of the invariant Quot-scheme and a moduli space construction such that:*

1. *The μ_θ-Harder-Narasimhan filtration of $\mathcal{F}$ is a subfiltration of its μ_D-Harder-Narasimhan filtration.*
2. *Moreover, the μ_D-Harder-Narasimhan filtration of $\mathcal{F}$ is a subfiltration of some μ_θ-Jordan-Hölder-Harder-Narasimhan filtration (i.e., a refinement of the semistable factors of the Harder-Narasimhan filtration by stable ones).*
3. *Even though the filtrations are never equal, the D-polygon converges to the θ-polygon, when $D \to \text{Irr } G$ as a directed set.*

In Fig. 5.2 it is illustrated how the filtrations and their associated polygons behave and converge for D large, as Theorem 5.6 establishes.

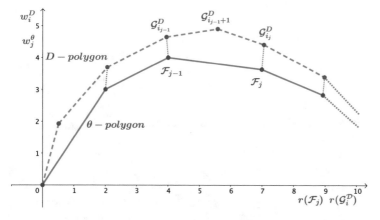

Fig. 5.2 θ-polygon and D-polygon of a (G, h)-constellation

Chapter 6
Stratifications on the Moduli Space of Higgs Bundles

One of the main utilities of the Harder-Narasimhan filtration is to provide a way to classify unstable objects of the same invariants (i.e., the same Harder-Narasimhan type) into geometric structures close to the idea of moduli spaces; the moduli of semistable objects collecting those with trivial Harder-Narasimhan filtration. This leads to stratifications of the moduli problem into regions where all objects of each locus share the same invariants in their Harder-Narasimhan filtrations (see, e.g., the works [40, 43, 49, 59, 83]). In this final chapter we review certain results on stratifications of the moduli space of Higgs bundles, performed with the invariants provided by the Harder-Narasimhan filtrations.

The moduli space of Higgs bundles has two stratifications. The Shatz stratification arises from the Harder-Narasimhan type of the underlying vector bundle of the Higgs bundle, and the Białynicki-Birula stratification comes from action of the nonzero complex numbers by multiplication on the Higgs field and Morse theory. While these two stratifications coincide in the case of rank two Higgs bundles, this is not the case in higher rank. We thus analyze the relationship between the two stratifications for the moduli space of rank three Higgs bundles, based on results contained in [31, 98].

The study of the whole moduli problem, containing the unstable locus, has granted many results for the topology of Higgs bundles moduli spaces (see [2, 22, 30, 40–42, 63, 77]). In the final section we use the correspondence between these two stratifications and the Hodge bundles defined by them, to provide certain information about the homotopy groups of the moduli space of k-Higgs bundles, where the Higgs field has poles of order k. It turns out that, under some conditions, the homotopy groups π_j stabilize when the index j grows, as it can be seen in [98–100].

© The Author(s), under exclusive license to Springer Nature Switzerland AG 2021
A. Zamora Saiz, R. A. Zúñiga-Rojas, *Geometric Invariant Theory, Holomorphic Vector Bundles and the Harder-Narasimhan Filtration*, SpringerBriefs in Mathematics, https://doi.org/10.1007/978-3-030-67829-6_6

6.1   Shatz Stratification

Let X be a compact Riemann surface and let $\mathcal{M}^H(r, d)$ be the moduli space of semistable Higgs bundles (E, φ), where E is a holomorphic vector bundle of rank r and degree d over X, $\varphi : E \to E \otimes \omega_X$ is the Higgs field and ω_X is the canonical bundle over X (see Sect. 4.4). Recall from Sect. 4.2 that any holomorphic vector bundle over X has a unique Harder-Narasimhan filtration:

$$0 = E_0 \subsetneq E_1 \subsetneq E_2 \subsetneq \cdots \subsetneq E_t \subsetneq E_{t+1} = E$$

verifying the two properties of Theorem 4.1. The Harder-Narasimhan type

$$\overrightarrow{\mu}(E) = \left(\mu(E^1), \ldots, \mu(E^1), \mu(E^2), \ldots, \mu(E^2), \ldots, \mu(E^{t+1}), \ldots, \mu(E^{t+1}) \right)$$

is the sequence of slopes of the quotients $E^i := E_i/E_{i-1}$ in the filtration, each repeated $\mathrm{rk}(E^i)$ times.

As a consequence of the work of Shatz [83, Propositions 10 and 11], there is a finite stratification of $\mathcal{M}^H(r, d)$ by the Harder-Narasimhan type of the underlying vector bundle E:

$$\mathcal{M}^H(r, d) = \bigcup_{\overrightarrow{\mu}(E)} U'_{\overrightarrow{\mu}(E)} ,$$

where $U'_{\overrightarrow{\mu}(E)} \subset \mathcal{M}^H(r, d)$ is the subspace of Higgs bundles (E, φ) whose underlying vector bundle E has Harder-Narasimhan type $\overrightarrow{\mu}(E)$. This stratification is known as the *Shatz stratification*. Note that there is an open dense stratum corresponding to Higgs bundles (E, φ) such that the underlying vector bundle E is itself stable, that is, its Harder-Narasimhan filtration is trivial and its Harder-Narasimhan type is

$$\overrightarrow{\mu}(E) = \left(\frac{d}{r}, \ldots, \frac{d}{r} \right) ,$$

the same slope repeated a number of times equal to $\mathrm{rk}(E)$. Since

$$\varphi \in H^0(X, \mathrm{End}(E) \otimes \omega_X) \simeq H^1(X, \mathrm{End}(E))^\vee$$

by Serre duality, and then understanding how the data of a Higgs field is recovered by elements of the cotangent bundle of X, stable Higgs bundles represent points in the cotangent bundle of the moduli space of stable holomorphic vector bundles $\mathcal{M}_s(r, d)$, namely

$$U'_{(d/r, \ldots, d/r)} = \left(T\mathcal{M}_s(r, d) \right)^\vee \subset \mathcal{M}^H(r, d) .$$

6.2 $\mathbb{C}^*$-Action and Białynicki-Birula Stratification

A Kähler manifold is a symplectic variety (X, ω) with an integrable almost-complex structure J such that the bilinear form

$$g(u, v) = \omega(u, Jv)$$

is a Riemannian metric. Under these conditions, the form ω is a real $(1, 1)$-form which is closed, that is, $d\omega = 0$, and is called the *Kähler form*. This way, Kähler manifolds are spaces where complex, symplectic, and Riemannian geometry interplay in a compatible way.

The moduli space of semistable Higgs bundles $\mathcal{M}^H(r, d)$ is a Kähler manifold (c.f. [45]) endowed with an action of the group $\mathbb{C}^*$ defined by multiplication (see [86]):

$$z \cdot (E, \varphi) \longmapsto (E, z \cdot \varphi) .$$

This action is biholomorphic with respect to the complex structure J, and the Kähler form ω is invariant under the restriction of the action to the circle $S^1 \subset \mathbb{C}^*$:

$$e^{i\theta} \cdot (E, \varphi) = (E, e^{i\theta} \cdot \varphi) .$$

This circle action is Hamiltonian (c.f. Sect. 3.3) with moment map given by:

$$\psi : \mathcal{M}^H(r, d) \longrightarrow \mathbb{R} , \quad \psi(E, \varphi) = \frac{1}{2\pi} \|\varphi\|_{L^2}^2 = \frac{i}{2\pi} \int_X \mathrm{tr}(\varphi\varphi^*) ,$$

where φ^* is the adjoint of φ with respect to the hermitian metric on E.

The moment map ψ is proper and has finitely many critical values, which are exactly the fixed points of the circle action on the moduli space. The fixed points of the form (E, φ) with $\varphi = 0$ correspond to the critical value $c_0 := 0$, and the associated critical submanifold is:

$$F_0 = \psi^{-1}(c_0) = \psi^{-1}(0) = \mathcal{M}(r, d) ,$$

which equals the moduli space of semistable vector bundles [13, 45, 86]. If $\varphi \neq 0$, the fixed points of the circle action are given by *variations of the Hodge structure* (c.f. [86]) which are *Hodge bundles*

$$E = E_1 \oplus E_2 \oplus \cdots \oplus E_p$$

with the Higgs field verifying

$$\varphi : E_j \longrightarrow E_{j+1} \otimes \omega_X , \quad 1 \leq j \leq p - 1 ,$$

and *type* $\big(\mathrm{rk}(E_1), \ldots, \mathrm{rk}(E_p)\big)$, the sequence of ranks of the bundle decomposition of (E, φ). The critical values c_λ corresponding to these critical points, where λ denotes the index of the critical point for the Morse-Bott function ψ, will depend on the degrees d_j of the subbundles E_j, $1 \leq j \leq p$.

By Morse theory, there exists a stratification of $\mathcal{M}^H(r, d)$ by critical irreducible submanifolds $F_\lambda = \psi^{-1}(c_\lambda)$, for each nonzero critical value c_λ. Consider the *upper flow sets*

$$U_\lambda^+ := \left\{ (E, \varphi) \in \mathcal{M}^H(r, d) : \lim_{z \to 0} z \cdot (E, \varphi) \in F_\lambda \right\}$$

of Higgs bundles whose limits under the $\mathbb{C}^*$-action lie on F_λ. Then we have the *Białynicki-Birula stratification* of $\mathcal{M}^H(r, d)$ (cf. [8]):

$$\mathcal{M}^H(r, d) = \bigcup_\lambda U_\lambda^+ ,$$

indexed by the critical values of ψ.

The distinguished component of the fixed locus $F_0 = \mathcal{M}(r, d)$ equaling the moduli space of semistable vector bundles corresponds to the minimal Białynicki-Birula stratum U_0^+. The Higgs bundles whose limit under the $\mathbb{C}^*$-action live in F_0 are precisely the semistable Higgs bundles whose underlying vector bundle is also semistable (c.f. [31, Proposition 2.2], and hence, we have the identification between the lower stratum in both Shatz and Białynicki-Birula stratifications:

$$U'_{(d/r, \ldots, d/r)} = U_0^+ .$$

Consider a semistable rank two and degree d Higgs bundle (E, φ), which corresponds to a fixed point by the $\mathbb{C}^*$-action on the moduli space $\mathcal{M}^H(2, d)$. If the underlying E is stable, it is a Hodge bundle of the form $(E, 0)$. If not, (E, φ) is a Hodge bundle of the form

$$\left(E = E_1 \oplus E/E_1, \quad \varphi = \begin{pmatrix} 0 & 0 \\ \varphi_{21} & 0 \end{pmatrix} \right)$$

where, if the underlying E is semistable, then $\mu(E_1) = \mu(E/E_1) = \mu(E) = \frac{d}{2}$ and $\varphi_{21} = 0$. If E is unstable, otherwise, $\mu(E_1) > \mu(E/E_1)$ and $\varphi_{21} \neq 0$.

Hausel shows (c.f. [40, Proposition 4.3.2]) that the Shatz stratification of the moduli space of Higgs bundles $\mathcal{M}^H(2, d)$ into Harder-Narasimhan types of the underlying bundle E coincides with the Białynicki-Birula stratification of the same moduli space. The strata corresponding to a semistable underlying E are

$$U'_{(d/2, d/2)} = U_0^+ ,$$

and the strata with different Harder-Narasimhan types, which for rank 2 bundles with Harder-Narasimhan filtration $0 \subsetneq E_1 \subsetneq E$ are of the form

$$\vec{\mu}(E) = (d_1, d - d_1) , \quad \text{with } d = \deg(E) , \ d_1 = \deg(E_1) ,$$

become identified as

$$U'_{(d_1, d-d_1)} = U^+_{d_1} ,$$

the Białynicki-Birula stratification being indexed by the degree of E_1. Moreover, for d odd and rank two there are no strictly semistable vector bundles, then the lower stratum is precisely the cotangent bundle of the moduli space of vector bundles:

$$U^+_0 = U'_{(d/2, d/2)} = (T M_s(2, d))^\vee = (T M(2, d))^\vee \subset \mathcal{M}^H(2, d) .$$

6.3 Stratifications in Rank Three

Motivated by the results of Hausel [40] on stratifications of the moduli space of rank two Higgs bundles, there has been some developments about stratifications for rank three. The results of this section can be found in [31].

First, we present bounds on Harder-Narasimhan types for a rank three Higgs bundle (E, φ). In general, the Harder-Narasimhan filtration of its underlying vector bundle E can be written as

$$0 = E_0 \subseteq E_1 \subseteq E_2 \subseteq E_3 = E$$

with $\vec{\mu}(E) = (\mu^1, \mu^2, \mu^3)$ the Harder-Narasimhan type of the underlying bundle E, where $\mu^i := \mu(E^i)$ and $E^i := E_i/E_{i-1}, i = 1, 2, 3$. The Harder-Narasimhan type verifies

$$\mu^1 \geq \mu^2 \geq \mu^3$$

and, with this notation, we allow shorter filtrations with $E_i = E_{i+1}$ if $\mu^i = \mu^{i+1}$. In the case that $\mu^1 = \mu^2 > \mu^3$, the Harder-Narasimhan filtration is

$$0 = E_0 \subsetneq E_1 = E_2 \subsetneq E_3 = E$$

and $\text{rk}(E_1) = \text{rk}(E_2) = 2$. Similarly, if $\mu^1 > \mu^2 = \mu^3$ we have $\text{rk}(E_1) = 1$, $\text{rk}(E_2) = 3$, and $E_2 = E_3 = E$. In any case we have the relationship

$$\frac{\mu^1 + \mu^2 + \mu^3}{3} = \mu(E) .$$

Define the morphisms

$$\varphi_{21}\colon E_1 \longrightarrow E/E_1 \otimes \omega_X \quad \text{and} \quad \varphi_{32}\colon E_2 \longrightarrow E/E_2 \otimes \omega_X$$

induced by φ, and let

$$I = \varphi_{21}(E_1) \otimes \omega_X^{-1} \subset E/E_1 \quad \text{and} \quad N = \mathrm{Ker}(\varphi_{32}) \subset E_2$$

be the corresponding image and kernel subbundles of φ_{21} and φ_{32}, respectively. To be precise, $\varphi_{21}(E_1) \otimes \omega_X^{-1}$ is not a subbundle in general, then I has to be defined as the *saturated sheaf* of that image, which is the smallest subbundle of E/E_1 containing $\varphi_{21}(E_1) \otimes \omega_X^{-1}$.

Observe that, if (E, φ) is a stable Higgs bundle such that the underlying E is an unstable vector bundle of Harder-Narasimhan type $\vec{\mu}(E) = (\mu^1, \mu^2, \mu^3)$, then $E_1 \subset E$ is destabilizing (as a vector subbundle), and hence, by stability of the Higgs bundle (E, φ) we have $\varphi_{21} \neq 0$. Indeed, otherwise, we would have a subbundle $E_1 \subset E$ preserved by the Higgs field φ (i.e., if $\varphi_{21} = 0$ then $\varphi(E_1) \subset E_1 \otimes \omega_X$) with greater slope $\mu^1 = \mu(E_1) > \mu(E)$, therefore destabilizing (E, φ) as a Higgs bundle. Similarly, $E_2 \subset E$ is destabilizing as a vector subbundle and thus $\varphi_{32} \neq 0$ (unless $\mu^2 = \mu^3$, equivalent to $E_2 = E$).

A nonzero map between semistable bundles $E \to F$ verifies $\mu(E) < \mu(F)$; otherwise, the kernel or the image of this map can be shown to contradict the stability of E or F. If $\mu^1 > \mu^2$, then $\mathrm{rk}(E_1) = 1$ and $I \subset E/E_1$ is a line bundle, since $\varphi_{21} \neq 0$. Then, φ_{21} provides a nonzero map of line bundles $E_1 \to I \otimes \omega_X$ such that

$$\mu^1 = \mu(E_1) \leq \mu(I) + 2g - 2 \,,$$

where recall that the degree of the canonical bundle is $\deg(\omega_X) = 2g - 2$, g being the genus of X, and that the degree of the product of line bundles is the sum of their degrees. Besides this, by definition of the Harder-Narasimhan filtration, since $E_2/E_1 \subset E/E_1$ is the maximal destabilizing subbundle any other subbundle like $I \subset E/E_1$ has lower slope, then

$$\mu(I) \leq \mu(E_2/E_1) = \mu^2 \,.$$

On the other hand, if $\mu^2 > \mu^3$, then it is $\mathrm{rk}(E_2) = 2$ and $N \subset E_2$ is a line bundle since $\varphi_{32} \neq 0$. Then, the image of φ_{32} has rank 1 and the kernel has also rank 1. Again, we get a nonzero map of line bundles

$$E_2/N \longrightarrow E/E_2 \otimes \omega_X$$

and we can compute

$$\mu(E_2/N) = \frac{\deg(E_2) - \deg(N)}{\mathrm{rk}(E_2) - \mathrm{rk}(N)} = \deg(E_2) - \mu(N) = \mu^1 + \mu^2 - \mu(N) \,,$$

hence

$$\mu\left(E_2/N\right) \le \mu\left(E/E_2 \otimes \omega_X\right) = \mu\left(E/E_2\right) + 2g - 2 = \mu^3 + 2g - 2 \, .$$

Following the same idea, since $E_1 \subset E_2$ is maximal destabilizing, the subbundle N has slope

$$\mu(N) \le \mu(E_1) = \mu^1 \, .$$

Combining all these inequalities we can state (c.f. [31, Propositions 4.2 and 4.3]) the following bounds for the difference of slopes $\mu^i - \mu^{i+1}$, $i = 1, 2$, and the slopes of I and N:

$$0 \le \mu^1 - \mu^2 \le 2g - 2 \tag{6.1}$$

$$0 \le \mu^2 - \mu^3 \le 2g - 2 \tag{6.2}$$

$$\mu^1 - (2g - 2) \le \mu(I) \le \mu^2 \tag{6.3}$$

$$\mu^1 + \mu^2 - \mu^3 - (2g - 2) \le \mu(N) \le \mu^1 \tag{6.4}$$

The purpose now will be to analyze the limit of the $\mathbb{C}^*$-action as $z \to 0$ in terms of the Harder-Narasimhan type of the underlying bundle E of a Higgs bundle (E, φ). Assume, from now on, that we are in the coprime case $\mathrm{GCD}(3, d) = 1$, where there are no strictly semistable bundles and, then

$$\mathcal{M}^H(3, d) = \mathcal{M}_s^H(3, d) \, .$$

Note that, because of the coprimality condition, none of the slopes μ^i is equal to $\mu(E)$; in particular $\mu^2 \ne \mu(E)$.

It is shown in [31, Proposition 2.2] that, if (E, φ) is a stable Higgs bundle, the limit is

$$\lim_{z \to 0} z \cdot (E, \varphi) = (E, 0)$$

if and only if E is stable. Now, let us check on nontrivial Harder-Narasimhan filtrations and consider rank three semistable Higgs bundles (E, φ) whose underlying vector bundle is unstable. The following result classifies the limit point under the $\mathbb{C}^*$-action, depending on the numerical invariants of E and the subbundles I and N previously defined.

Theorem 6.1 ([31, Theorem 5.1]) *Let (E, φ) be a stable Higgs bundle in $\mathcal{M}^H(3, d)$ (with $\mathrm{GCD}(3, d) = 1$), such that the underlying E is an unstable vector bundle of slope $\mu(E)$ and Harder-Narasimhan type $\overrightarrow{\mu}(E) = \left(\mu^1, \mu^2, \mu^3\right)$.*

Define I and N as before, and let $Q := (E/E_1)/I$. *Then, the limit* $(E_0, \varphi_0) = \lim_{z \to 0} (E, z \cdot \varphi)$ *falls into one of the following cases:*

(1) *If* $\mu^2 < \mu(E)$. *Then* $\mu^1 > \mu^2 \geq \mu^3$ *and one of the following alternatives holds.*

　(1.1) *Either* $\mu^1 - (2g - 2) \leq \mu(I) < P$, *and* (E_0, φ_0) *is the following type* (1, 2)-*Hodge bundle:*

$$(E_0, \varphi_0) = \left(E_1 \oplus E/E_1, \left(\begin{smallmatrix} 0 & 0 \\ \varphi_{21} & 0 \end{smallmatrix} \right) \right) ,$$

　　with Harder-Narasimhan type $\overrightarrow{\mu}(E_0) = (\mu^1, \mu^2, \mu^3) = \overrightarrow{\mu}(E)$ *and associated graded vector bundle* $\mathrm{gr}(E_0) = \mathrm{gr}(E)$.

　(1.2) *Or* $P < \mu(I) \leq \mu^3$, *and* (E_0, φ_0) *is the following type* (1, 1, 1)-*Hodge bundle:*

$$(E_0, \varphi_0) = \left(E_1 \oplus I \oplus Q, \left(\begin{smallmatrix} 0 & 0 & 0 \\ \varphi_{21} & 0 & 0 \\ 0 & \varphi_{32} & 0 \end{smallmatrix} \right) \right) ,$$

　　with Harder-Narasimhan type $\overrightarrow{\mu}(E_0) = (\mu^1, \mu(Q), \mu(I))$ *and associated graded vector bundle* $\mathrm{gr}(E_0) = E_1 \oplus Q \oplus I$.

　(1.3) *Or* $\mu(I) = \mu^2$, *then it is necessarily* $\mu^3 < \mu^2$, $I = E_2/E_1$ *and* $Q = E/E_2$, *and* (E_0, φ_0) *is the following type* (1, 1, 1)-*Hodge bundle:*

$$(E_0, \varphi_0) = \left(E_1 \oplus E_2/E_1 \oplus E/E_2, \left(\begin{smallmatrix} 0 & 0 & 0 \\ \varphi_{21} & 0 & 0 \\ 0 & \varphi_{32} & 0 \end{smallmatrix} \right) \right) ,$$

　　with Harder-Narasimhan type $\overrightarrow{\mu}(E_0) = (\mu^1, \mu^2, \mu^3) = \overrightarrow{\mu}(E)$ *and associated graded vector bundle* $\mathrm{gr}(E_0) = \mathrm{gr}(E)$.

(2) *If* $\mu^2 > \mu(E)$. *Then* $\mu^1 \geq \mu^2 > \mu^3$ *and one of the following alternatives holds.*

　(2.1) *Either* $\mu^1 + \mu^2 - \mu^3 - (2g - 2) \leq \mu(N) < \mu(E)$, *and* (E_0, φ_0) *is the following type* (2, 1)-*Hodge bundle:*

$$(E_0, \varphi_0) = \left(E_2 \oplus E/E_2, \left(\begin{smallmatrix} 0 & 0 \\ \varphi_{32} & 0 \end{smallmatrix} \right) \right) ,$$

　　with $\overrightarrow{\mu}(E_0) = \overrightarrow{\mu}(E)$ *and* $\mathrm{gr}(E_0) = \mathrm{gr}(E)$.

　(2.2) *Or* $\mu(E) < \mu(N) \leq \mu^2$, *and* (E_0, φ_0) *is the following type* (1, 1, 1)-*Hodge bundle:*

$$(E_0, \varphi_0) = \left(N \oplus E_2/N \oplus E/E_2, \left(\begin{smallmatrix} 0 & 0 & 0 \\ \varphi_{21} & 0 & 0 \\ 0 & \varphi_{32} & 0 \end{smallmatrix} \right) \right) ,$$

with $\vec{\mu}(E_0) = (\mu^1 + \mu^2 - \mu(N), \mu(N), \mu^3)$ and $\mathrm{gr}(E_0) = E_2/N \oplus N \oplus E/E_2$.

(2.3) Or $\mu(N) = \mu^1$, then the strict inequality $\mu^1 > \mu^2$ holds with $N = E_1$, and (E_0, φ_0) is the following type $(1, 1, 1)$-Hodge bundle:

$$(E_0, \varphi_0) = \left(E_1 \oplus E_2/E_1 \oplus E/E_2, \begin{pmatrix} 0 & 0 & 0 \\ \varphi_{21} & 0 & 0 \\ 0 & \varphi_{32} & 0 \end{pmatrix} \right) ,$$

with $\vec{\mu}(E_0) = \vec{\mu}(E)$ and $\mathrm{gr}(E_0) = \mathrm{gr}(E)$.

6.3.1 Sketch of the Proof of Theorem 6.1

We will provide here the proof of Case 1 in Theorem 6.1, Case 2 being similar (see [31]). It is, for sure, a good illustration of how to work with many of the objects discussed in the previous chapters: vector bundles, their morphisms, ranks, degrees and slopes, orbits, limits and quotient spaces, and above all of this, Harder-Narasimhan filtrations and Harder-Narasimhan polygons and types, one of the core ideas of this book.

For the proof, we rely on the complex analytic construction of the moduli space of Higgs bundles (see Sects. 4.3 and 4.4): let $\mathbb{E}$ be the smooth complex vector bundle underlying E over X and consider the pair

$$\left(\overline{\partial}_E, \varphi\right) \in \mathcal{A}^{0,1}(\mathbb{E}) \times \mathcal{E}_X^{1,0}(\mathrm{End}(\mathbb{E}))$$

representing (E, φ) in the configuration space of all Higgs bundles. Our strategy of proof is to find a family of complex gauge transformations $g(z) \in \mathcal{G}^{\mathbb{C}}$ parametrized by $z \in \mathbb{C}^*$, such that the limit in the configuration space

$$\left(\overline{\partial}_{E_0}, \varphi_0\right) = \lim_{z \to 0} \left(g(z) \cdot \left(\overline{\partial}_E, \varphi\right)\right) = \lim_{z \to 0} \left(\overline{\partial}_E, g(z) \cdot \varphi\right)$$

gives a *stable* Higgs bundle (E_0, φ_0). It will then follow that (E_0, φ_0) represents the limit in the moduli space $\mathcal{M}^H(3, d)$.

Recall the Harder-Narasimhan polygon of E (see Fig. 4.2) which, in the rank three case, is made of three segments of slopes $\mu^1 \geq \mu^2 \geq \mu^3$, and whose average is the slope of E

$$\frac{\mu^1 + \mu^2 + \mu^3}{3} = \mu(E) .$$

Suppose, as in the Case 1 of Theorem 6.1, that $\mu^2 < \mu(E)$. Then, because of the inequalities and the average condition, we must necessarily have $\mu^1 > \mu^2 \geq \mu^3$ (see Fig. 6.1).

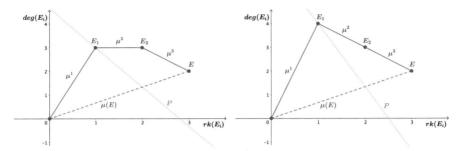

Fig. 6.1 Harder-Narasimhan polygon for the case $\mu^1 > \mu^2 \geq \mu^3$. On the left we represent the case $\mu^2 > \mu^3$ and on the right we have the case $\mu^2 = \mu^3$. Downwards line represents the slope P to compare with $\mu(I)$ in each case (see Fig. 6.2)

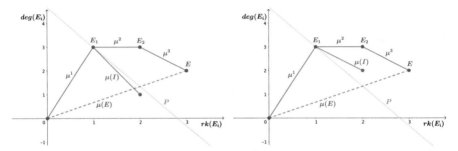

Fig. 6.2 Harder-Narasimhan polygon for the case $\mu^1 > \mu^2 > \mu^3$. On the left we represent the situation for $\mu(I) < P$ and on the right we represent the situation for $\mu(I) > P$. The case $\mu^2 = \mu^3$ is similar

It follows from (6.3) and the previous explanation that $I \subset E/E_1$ is a line bundle since $\varphi_{21} \neq 0$, and that

$$\mu^1 - (2g - 2) \leq \mu(I) \leq \mu^2 .$$

The line bundle I is a subbundle of E/E_1, hence in the corresponding Harder-Narasimhan polygon the slope $\mu(I)$ can be represented as a segment starting at the point with coordinates $(\mathrm{rk}(E_1), \deg(E_1)) = (1, \mu^1)$, as in Fig. 6.2. The position of this slope with respect to the rest will determine the role played by I in the Harder-Narasimhan filtration of the limit E_0. Because of this, we define

$$P := -\frac{1}{3}\mu^1 + \frac{2}{3}\mu^2 + \frac{2}{3}\mu^3 = -\mu^1 + 2\mu(E) ,$$

and note that the conditions $\mu(I) < P$ and $\mu(I) > P$ correspond to the two situations in Fig. 6.2 where the segment of I in the Harder-Narasimhan polygon ends, respectively, below or above the segment corresponding to E, of slope $\mu(E)$. Also note that, since $\mathrm{GCD}(3, d) = 1$, it cannot be $\mu(I) = P$, therefore we only discuss the other two strict inequality cases.

- **Case A:** $\mu^1 - (2g - 2) \leq \mu(I) < P$.

  From $E_1 \subset E$, we can construct the short exact sequence:

  $$0 \longrightarrow E_1 \longrightarrow E \longrightarrow E/E_1 \longrightarrow 0 . \tag{6.5}$$

Note that an exact sequence of holomorphic bundles is not split in general, because there are not enough holomorphic functions to have a section $E/E_1 \rightarrow E$. However, for the underlying smooth complex bundles there is always a splitting, because of the existence of partitions of the unity to define a differentiable section (this is the same as saying that the sheaf $\mathcal{E}_X$ is fine, c.f. Sect. 2.2). Therefore, let $\mathbb{E}$, $\mathbb{E}_1$, and $\mathbb{E}_2$ be the smooth complex vector bundles underlying E, E_1, and E/E_1, respectively, and consider the splitting

$$\mathbb{E} \simeq \mathbb{E}_1 \oplus \mathbb{E}_2 . \tag{6.6}$$

With respect to this decomposition, the holomorphic structure on $\mathbb{E}$ is given by the $\bar{\partial}$-operator

$$\bar{\partial}_E = \begin{pmatrix} \bar{\partial}_1 & \beta \\ 0 & \bar{\partial}_2 \end{pmatrix} ,$$

where $\bar{\partial}_1$ and $\bar{\partial}_2$ are the differential operators defining the corresponding holomorphic structures on $\mathbb{E}_1$ and $\mathbb{E}_2$, respectively, and $\beta \in \mathcal{A}^{0,1}\left(\mathrm{Hom}(\mathbb{E}_2, \mathbb{E}_1)\right)$. Similarly, the Higgs field $\varphi \in \mathcal{E}^{1,0}\left(\mathrm{End}(\mathbb{E})\right)$ can be written as

$$\varphi = \begin{pmatrix} \varphi_{11} & \varphi_{12} \\ \varphi_{21} & \varphi_{22} \end{pmatrix} .$$

Define the complex gauge subgroup of transformations by

$$g(z) = \begin{pmatrix} 1 & 0 \\ 0 & z \end{pmatrix} \in \mathcal{G}^{\mathbb{C}} , \quad \text{with } z \in \mathbb{C}^* ,$$

also with respect to the decomposition (6.6). The action of $g(z)$ on $\left(\bar{\partial}_E, z \cdot \varphi\right)$ and the limits when z goes to 0 are given by

$$g(z) \cdot \bar{\partial}_E = g(z)^{-1} \circ \bar{\partial}_E \circ g(z) = \begin{pmatrix} \bar{\partial}_1 & z \cdot \beta \\ 0 & \bar{\partial}_2 \end{pmatrix} \xrightarrow{z \to 0} \begin{pmatrix} \bar{\partial}_1 & 0 \\ 0 & \bar{\partial}_2 \end{pmatrix} ,$$

and

$$g(z) \cdot (z \cdot \varphi) = g(z)^{-1}(z \cdot \varphi)g(z) = \begin{pmatrix} z \cdot \varphi_{11} & z^2 \cdot \varphi_{12} \\ \varphi_{21} & z \cdot \varphi_{22} \end{pmatrix} \xrightarrow{z \to 0} \begin{pmatrix} 0 & 0 \\ \varphi_{21} & 0 \end{pmatrix} .$$

Observe that this formula for the gauge transformed $\bar{\partial}$-operator is valid because the gauge transformation is constant on the base Riemann surface X, affecting just the holomorphic structure on the bundle.

Thus, in the configuration space $\mathcal{A}^{0,1}(\mathbb{E}) \times \mathcal{E}_X^{1,0}(\mathrm{End}(\mathbb{E}))$ we get that the limit $\lim_{z \to 0} z \cdot (E, \varphi)$ is gauge equivalent to

$$(E_0, \varphi_0) = \left(E_1 \oplus E/E_1, \begin{pmatrix} 0 & 0 \\ \varphi_{21} & 0 \end{pmatrix} \right).$$

Now we show that this (E_0, φ_0) is stable, therefore it defines a point in the moduli space $\mathcal{M}^H(3, d)$. To do this, we need to find the subbundles of E_0 which are φ_0-invariant. This can be done by looking at the matrix of φ_0, adapted to the decomposition (6.6), where we see that these come from E/E_1 and line subbundles $L \subset E/E_1$, but also from $E_1 \oplus I$ by definition of I in terms of φ_{21}. We will show that the slopes of these three subbundles are smaller than $\mu(E_0) = \mu(E)$, hence they do not contradict the stability of (E_0, φ_0) as a Higgs bundle. Figure 6.3 represents the three cases, with the corresponding segments translated to the origin to compare the slopes with $\mu(E)$.

- The rank 2 subbundle $E_1 \oplus I \subset E_1 \oplus E/E_1 = E_0$ has slope

$$\mu(E_1 \oplus I) = \frac{\deg(E_1) + \deg(I)}{2} = \frac{\mu^1 + \mu(I)}{2} < \frac{\mu^1 + P}{2} = \mu(E) = \mu(E_0),$$

using that $\mu(I) < P$.

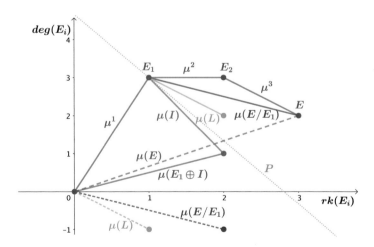

Fig. 6.3 Harder-Narasimhan polygon for the case $\mu^1 > \mu^2 > \mu^3$ and $\mu(I) < P$. Subbundles $E_1 \oplus I$, E/E_1, and L are plotted and observed to have slope less than $\mu(E)$. The case $\mu^2 = \mu^3$ is similar

- The slope of the rank 2 subbundle $E/E_1 \subset E_1 \oplus E/E_1 = E_0$ is the average of μ^2 and μ^3. Given that $\mu(E) > \mu^2 \geq \mu^3$ then $\mu(E/E_1) < \mu(E) = \mu(E_0)$.
- For the line subbundle $L \subset E/E_1$ we do the following. Either E/E_1 is semistable, or its maximal destabilizer is E_2/E_1, as in Fig. 6.3, then no line subbundle has a slope above μ^2. Hence $\mu(L) \leq \mu^2 < \mu(E) = \mu(E_0)$.

Finally, note that the Harder-Narasimhan filtration of the vector bundle in the limit, $E_0 = E \oplus E/E_1$, has the same factors as the one of E:

$$\text{either } 0 \subset E_1 \subset E_2 \subset E \quad \text{if } \mu^2 > \mu^3 \, ,$$

$$\text{or } 0 \subset E_1 \subset E \quad \text{if } \mu^2 = \mu^3 \, .$$

Therefore the graded objects of E and E_0 coincide, completing Case (1.1).

- **Case B:** $P < \mu(I) \leq \mu^2$.

We now define $Q := (E/E_1)/I$ and construct a short exact sequence as before

$$0 \longrightarrow I \longrightarrow E/E_1 \longrightarrow Q \longrightarrow 0 \, . \tag{6.7}$$

Combining the two sequences (6.5) and (6.7), we obtain the splitting

$$\mathbb{E} = \mathbb{E}_1 \oplus \mathbb{I} \oplus \mathbb{Q} \, , \tag{6.8}$$

where $\mathbb{E}_1$, $\mathbb{I}$, and $\mathbb{Q}$ are the smooth complex vector bundles underlying E_1, I, and Q, respectively. Adapted to this decomposition, the holomorphic structure on E takes the form

$$\bar{\partial}_E = \begin{pmatrix} \bar{\partial}_1 & \beta_{12} & \beta_{13} \\ 0 & \bar{\partial}_2 & \beta_{23} \\ 0 & 0 & \bar{\partial}_3 \end{pmatrix} \, ,$$

and the Higgs field φ is

$$\varphi = \begin{pmatrix} \varphi_{11} & \varphi_{12} & \varphi_{13} \\ \varphi_{21} & \varphi_{22} & \varphi_{23} \\ 0 & \varphi_{32} & \varphi_{33} \end{pmatrix} \, .$$

The gauge transformation that we will use now is, for each $z \in \mathbb{C}^*$:

$$g(z) := \begin{pmatrix} 1 & 0 & 0 \\ 0 & z & 0 \\ 0 & 0 & z^2 \end{pmatrix}$$

acting on $(E, z \cdot \varphi)$ by

$$g(z) \cdot \bar{\partial}_E = g(z)^{-1} \circ \bar{\partial}_E \circ g(z) = \begin{pmatrix} \bar{\partial}_1 & z \cdot \beta_{12} & z^2 \cdot \beta_{13} \\ 0 & \bar{\partial}_2 & z \cdot \beta_{23} \\ 0 & 0 & \bar{\partial}_3 \end{pmatrix} \xrightarrow{z \to 0} \begin{pmatrix} \bar{\partial}_1 & 0 & 0 \\ 0 & \bar{\partial}_2 & 0 \\ 0 & 0 & \bar{\partial}_3 \end{pmatrix} \quad \text{and}$$

$$g(z) \cdot (z \cdot \varphi) = g(z)^{-1}(z \cdot \varphi) g(z) = \begin{pmatrix} z \cdot \varphi_{11} & z^2 \cdot \varphi_{12} & z^3 \cdot \varphi_{13} \\ \varphi_{21} & z \cdot \varphi_{22} & z^2 \cdot \varphi_{23} \\ 0 & \varphi_{32} & z \cdot \varphi_{33} \end{pmatrix} \xrightarrow{z \to 0} \begin{pmatrix} 0 & 0 & 0 \\ \varphi_{21} & 0 & 0 \\ 0 & \varphi_{32} & 0 \end{pmatrix},$$

with the limits as indicated. Hence, in the configuration space, the limit $\lim_{z \to 0} z \cdot (E, \varphi)$ now is gauge equivalent to

$$(E_0, \varphi_0) = \left(E_1 \oplus I \oplus Q, \begin{pmatrix} 0 & 0 & 0 \\ \varphi_{21} & 0 & 0 \\ 0 & \varphi_{32} & 0 \end{pmatrix} \right).$$

Let us show now that (E_0, φ_0) is a semistable Higgs bundle by proving that the only two φ_0-invariant subbundles of E_0, according to the Higgs field matrix, are $I \oplus Q$ and Q. Again, Figs. 6.4 and 6.5 represent these cases with the segments translated to the origin to compare with $\mu(E)$.

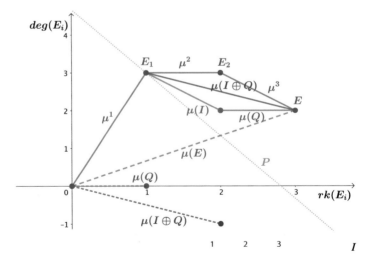

Fig. 6.4 Harder-Narasimhan polygon for the case $\mu^1 > \mu^2 > \mu^3$ and $P < \mu(I) \le \mu(Q)$. Subbundles I and $I \oplus Q$ are plotted and observed to have slope less than $\mu(E)$. The case $\mu^2 = \mu^3$ is similar with $\mu(I \oplus Q) = \mu^2 = \mu^3$

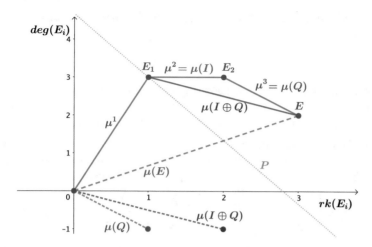

Fig. 6.5 Harder-Narasimhan polygon for the case $\mu^1 > \mu^2 > \mu^3$, $P < \mu(I)$, and $\mu(I) > \mu(Q)$. We see that $I = E_2/E_1$ and $Q = E/E_2$

- The slope of $I \oplus Q$ is

$$\mu\,(I \oplus Q) = \frac{\deg(I) + \deg(Q)}{2} = \frac{\deg(E) - \mu^1}{2} < \mu(E)\,,$$

where the last inequality is equivalent to $\mu^1 > \mu(E)$, which holds by the Harder-Narasimhan filtration decreasing slopes property.

- For Q the slope is

$$\mu(Q) = \frac{\deg(E) - \deg(E_1) - \deg(I)}{1} = 3\mu(E) - \mu^1 - \mu(I) < \mu(E)\,,$$

where the inequality is equivalent to $\mu(I) > P$.

Gathering the previous information, we distinguish two cases. First recall the result by Shatz [83, Theorem 3] which says that, after specialization, the Harder-Narasimhan polygon can only rise. This means that, as E_0 is a specialization of E (obtained as a limit), its Harder-Narasimhan polygon has to lie on or above the polygon of E. If $P < \mu(I) \leq \mu(Q)$ (as in Fig. 6.4), the Harder-Narasimhan type of E_0 is $\overrightarrow{\mu}(E_0) = (\mu(E_1), \mu(Q), \mu(I))$, then $\mu(I) \leq \mu^3$ and we conclude Case (1.2). If $\mu(I) > \mu(Q)$ the situation is represented in Fig. 6.5, and the Harder-Narasimhan type of E_0 is $\overrightarrow{\mu}(E_0) = (\mu(E_1), \mu(I), \mu(Q))$, and $\mu(I) \geq \mu^2$. But, since $I \subset E/E_1$, its slope cannot be greater than the slope of the maximal destabilizer, which is $\mu^2 = \mu(E_2/E_1)$ then $\mu(I) = \mu^2$, and moreover, $I = E_2/E_1$, by uniqueness of this maximal destabilizer. This implies $Q = E/E_2$ and completes Case (1.3). Note that by this Shatz specialization result, $\mu(I)$ cannot take any value strictly in between μ^3 and μ^2.

6.3.2 Relationship Between Shatz and Biłynicki-Birula Stratifications for Rank Three Higgs Bundles

From Theorem 6.1, we can deduce a couple of interesting consequences. The theorem shows that, in general, knowledge of the Harder-Narasimhan type of E does not suffice to determine the underlying bundle E_0 of the limit $(E_0, \varphi_0) = \lim_{z \to 0} (E, z \cdot \varphi)$. However, there are some Harder-Narasimhan types (μ^1, μ^2, μ^3) for which E_0 is, indeed, determined by E.

Observe that, by (6.1) and (6.2) one has $0 \leq \mu^1 - \mu^3 \leq 4g - 4$. Note also that in Cases (1.1) and (1.2) we have $\mu(I) \leq \mu^3$. Moreover, by (6.3) we have $\mu^1 - (2g - 2) \leq \mu(I)$. It follows that $\mu^1 - (2g - 2) \leq \mu^3$. Similarly, in Cases (2.1) and (2.2) we have $\mu(N) \leq \mu^2$ and, by (6.4), $\mu^1 + \mu^2 - \mu^3 - (2g - 2) \leq \mu(N)$. Hence, $\mu^1 + \mu^2 - \mu^3 - (2g - 2) \leq \mu^2$. Therefore, we have shown:

Corollary 6.1 ([31, Corollary 5.4]) *Let $(E, \varphi) \in \mathcal{M}^H(3, d)$ be such that E is an unstable vector bundle of slope μ and $\vec{\mu}(E) = (\mu^1, \mu^2, \mu^3)$. Assume that $\mu^1 - \mu^3 > 2g - 2$. Then the limit $(E_0, \varphi_0) = \lim_{z \to 0}(E, z \cdot \varphi)$ is given by Case (1.3) if $\mu^2 < \mu(E)$, or by Case (2.3) if $\mu^2 > \mu(E)$, in Theorem 6.1.*

In a similar vein, we shall next see that certain Hodge bundles (E_0, φ_0) can only be the limit of a Higgs bundle (E, φ) whose underlying vector bundle E has the same Harder-Narasimhan type as that of the Hodge bundle E_0. Recall that fixed points of type $(1, 1, 1)$ have the form

$$(E_0, \varphi_0) = \left(L_1 \oplus L_2 \oplus L_3, \begin{pmatrix} 0 & 0 & 0 \\ \varphi_{21} & 0 & 0 \\ 0 & \varphi_{32} & 0 \end{pmatrix} \right)$$

and can be parametrized by the invariants

$$m_1 = \deg(L_2) - \deg(L_1) + 2g - 2 \,,$$
$$m_2 = \deg(L_3) - \deg(L_2) + 2g - 2 \,,$$

subject to the conditions (see Gothen [30]):

$$m_i \geq 0 \,, \ i = 1, 2 \,,$$
$$2m_1 + m_2 < 6g - 6 \,,$$
$$m_1 + 2m_2 < 6g - 6 \,,$$
$$m_1 + 2m_2 \equiv 0 \,(\mathrm{mod}\ 3) \,.$$

Translating this to the degrees of L_i, the corresponding components $F_{(l_1, l_2, l_3)}$ of the fixed locus under the $\mathbb{C}^*$-action are indexed by tuples (l_1, l_2, l_3) where

$$l_i = \mu(L_i) = \deg(L_i) ,$$

subject to the constraints

$$l_{i+1} - l_i + 2g - 2 \geq 0 , \; i = 1, 2 ,$$

$$\frac{1}{3}l_1 + \frac{1}{3}l_2 - \frac{2}{3}l_3 > 0 ,$$

$$\frac{2}{3}l_1 - \frac{1}{3}l_2 - \frac{1}{3}l_3 > 0 .$$

On the one hand, (E_0, φ_0) being polystable and the condition $l_1 - l_3 > 2g - 2$, together imply that φ_{21} and φ_{32} are nonzero. On the other hand, from cases (1.3) and (2.3) in Theorem 6.1 we get the limit of a $(1, 1, 1)$-Hodge bundle with $\mu(L_1) - \mu(L_3) > 2g - 2$, and in these cases it is $E_0 \simeq \mathrm{gr}(E_0) = \mathrm{gr}(E)$. This shows the content of the following result:

Corollary 6.2 ([31, Corollary 5.5]) *Let*

$$(E_0, \varphi_0) = \left(L_1 \oplus L_2 \oplus L_3, \begin{pmatrix} 0 & 0 & 0 \\ \varphi_{21} & 0 & 0 \\ 0 & \varphi_{32} & 0 \end{pmatrix} \right)$$

be a Hodge bundle of type $(1, 1, 1)$ *with* $l_1 - l_3 > 2g - 2$. *Then* $l_1 > l_2 > l_3$ *and any Higgs bundle* (E, φ) *having* (E_0, φ_0) *as a limit has the same Harder-Narasimhan type and associated graded vector bundle.*

Corollaries 6.1 and 6.2 lead to an identification between some of the Shatz strata and Białynicki-Birula strata. Recall that the set $U'_{(l_1, l_2, l_3)}$ denotes the Shatz stratum of Higgs bundles whose Harder-Narasimhan type (of the underlying bundle E) is (l_1, l_2, l_3) and $U^+_{(l_1, l_2, l_3)}$ denotes the Białynicki-Birula stratum of Higgs bundles whose limits lie on the irreducible component $F_{(l_1, l_2, l_3)}$.

Theorem 6.2 ([31, Theorem 5.6]) *Let* (l_1, l_2, l_3) *be such that* $l_1 - l_3 > 2g - 2$. *Then the corresponding Shatz and Białynicki-Birula strata in the moduli space of Higgs bundles* $\mathcal{M}^H(3, d)$ *coincide:*

$$U'_{(l_1, l_2, l_3)} = U^+_{(l_1, l_2, l_3)} .$$

6.4 Homotopy Groups

The work of Hausel [40] proves that the Shatz stratification of the moduli space of rank two Higgs bundles coincides with its Białynicki-Birula stratification. One of the applications of this correspondence is to estimate some homotopy groups of the moduli spaces of k-Higgs bundles of rank two. We begin with the definition of these objects.

Let X be a compact Riemann surface. Fix a point $p \in X$ and let $O_X(p)$ be the associated line bundle to the degree 1 divisor $p \in X$ (see Sect. 2.3.3). Observe that the sections of $O_X(p)$ are meromorphic functions with a pole of order 1 at p. Similarly, the sections of the twist $O_X(kp)$ have poles of order k at the point p.

A k-*Higgs bundle* (also called a *Higgs bundle with poles of order k*) is a pair (E, φ^k) where the k-*Higgs field* is the morphism

$$\varphi^k \in H^0(X, \mathrm{End}(E) \otimes \omega_X(kp)) \ , \quad \varphi^k : E \longrightarrow E \otimes \omega_X \otimes O_X(kp) = E \otimes \omega_X(kp)$$

which is a Higgs field with poles of order k. The moduli space of semistable k-Higgs bundles of rank r and degree d is denoted by $\mathcal{M}^{H,k}(r, d)$. If we assume that $\mathrm{GCD}(r, d) = 1$, $\mathcal{M}^{H,k}(r, d)$ is a smooth projective variety.

We define an embedding

$$i_k : \mathcal{M}^{H,k}(r, d) \longrightarrow \mathcal{M}^{H,k+1}(r, d)$$
$$[(E, \varphi^k)] \longmapsto [(E, \varphi^k \otimes s_p)]$$

where $0 \neq s_p \in H^0(X, O_X(kp))$ is a nonzero fixed section of the line bundle $O_X(kp)$. The goal is to show that the induced map in the homotopy groups

$$\pi_n(i_k) : \pi_n\left(\mathcal{M}^{H,k}(r, d)\right) \longrightarrow \pi_n\left(\mathcal{M}^{H,k+1}(r, d)\right)$$

stabilizes as $k \to \infty$.

For rank two, the homotopy groups $\pi_n\left(\mathcal{M}^k(2, 1)\right)$ are estimated in [40] by using the coincidence between the Shatz and the Białynicki-Birula stratifications, plus the embeddings $\mathcal{M}^{H,k}(2, 1) \hookrightarrow \mathcal{M}^{H,k+1}(2, 1)$. Calling

$$\mathcal{M}^\infty(r, d) := \lim_{k\to\infty} \mathcal{M}^{H,k}(r, d) = \bigcup_{k=0}^{\infty} \mathcal{M}^{H,k}(r, d)$$

to the direct limit of the moduli spaces $\left\{\mathcal{M}^{H,k}(r, d)\right\}_{k=0}^{\infty}$, Hausel and Thaddeus [42] prove that the classifying space of $\overline{G}$, which is the quotient of the gauge unitary transformations by its constant central subgroup $U(1)$, is homotopically equivalent to the direct limit of the spaces $\mathcal{M}^{H,k}(r, d)$:

$$B\overline{G} \simeq \mathcal{M}^\infty(r, d) = \lim_{k\to\infty} \mathcal{M}^{H,k}(r, d) \ .$$

From this, in the case of rank two and three, it can be shown that the inverse limits of the cohomology rings are:

$$\varprojlim H^* \left(\mathcal{M}^{H,k}(r,d) \,,\, \mathbb{Z} \right) \simeq H^* \left(\mathcal{M}^\infty(r,d) \,,\, \mathbb{Z} \right) \simeq H^* \left(B\overline{\mathcal{G}} \,,\, \mathbb{Z} \right) \,.$$

Unfortunately, for rank three the two stratifications of the moduli space of Higgs bundles do not coincide completely. In spite of this, in [99] some estimates for homotopy groups of the moduli spaces of k-Higgs bundles $\mathcal{M}^{H,k}(3,d)$ over a compact Riemann surface X of genus $g > 2$ are provided.

Fixing the determinant $\det(E) = \Lambda \in \mathrm{Jac}^d(X)$ in the degree d component of the Jacobian of line bundles on X, denote by $\mathcal{M}^{H,k}_\Lambda$ the moduli space of rank three k-Higgs bundles with fixed determinant Λ, and let

$$\mathcal{M}^\infty_\Lambda := \lim_{k \to \infty} \mathcal{M}^{H,k}_\Lambda$$

be the direct limit of these moduli spaces, as before. It is shown in [99] that, for the fixed determinant case, the group action of $\pi_1(\mathcal{M}^{H,k}_\Lambda)$ on the higher relative homotopy groups $\pi_n \left(\mathcal{M}^\infty_\Lambda, \mathcal{M}^{H,k}_\Lambda \right)$ is trivial. Using this, it is proved (c.f. [99, Corollary 4.14]) that for every n there exists a k_0, depending on n, such that there are isomorphisms for the lower homotopy groups

$$\pi_j \left(\mathcal{M}^{H,k}_\Lambda \right) \xrightarrow{\;\simeq\;} \pi_j \left(\mathcal{M}^\infty_\Lambda \right) \,,$$

for all $k \geq k_0$ and for all $j \leq n-1$.

References

1. Alexandrino M.M., Bettiol R.G., *Lie groups and Geometric Aspects of Isometric Actions,* Springer International Publishing Switzerland, (2015).
2. Álvarez-Cónsul L., García-Prada O., Schmitt A.H.W., *On the geometry of moduli spaces of holomorphic chains over compact Riemann surfaces,* IMRP Int. Math. Res. Pap. **82** Art. ID 73597, (2006).
3. Álvarez-Cónsul L., King A.D., *A functorial construction of moduli of sheaves,* Invent. Math. **168**, (2007) 613–666.
4. Atiyah M.F., *Vector bundles over an elliptic curve,* Proc. London. Math. Soc. **7**, (1957) 414–452.
5. Atiyah M.F., Bott R., *The Yang-Mills equations over Riemann surfaces,* Phil. Trans. R. Soc. Lond. **308** no. 1505, (1982) 523–615.
6. Atiyah M.F., MacDonald, I.G., *Introduction to commutative algebra,* Addison-Wesley, Reading, Mass., (1969).
7. Becker T., Terpereau R., *Moduli spaces of (G, h)-constellations,* Transform. Groups **20** (2), (2015) 335–366.
8. Białynicki-Birula A., *Some theorems on actions of algebraic groups,* Ann. of Math. **98**, (1973) 480–497.
9. Biswas I., Zamora A., *On the Gieseker Harder-Narasimhan filtration for principal bundles,* Bull. Sci. math **140** Issue 4, (2016) 58–69.
10. Borel, A., *Linear Algebraic Groups,* second edition, Grad. Texts in Math., **126**, Springer-Verlag New York, (1991).
11. Bott, R., Tu, L.W., *Differential Forms in Algebraic Topology,* Grad. Texts in Math. **82**, Springer-Verlag New York, (1982).
12. Bradlow, S.B., *Special metrics and stability for holomorphic bundles with global sections,* J. Diff. Geom. **33**, (1991) 169–214.
13. Bradlow, S.B., García-Prada, O., Gothen P.B., *Homotopy groups of moduli spaces of representations,* Topology **47**, (2008) 203–224.
14. Bredon G., *Introduction to Compact Transformation Groups,* Academic Press, (1972).
15. Bruasse, L. *Optimal destabilizing vectors in some gauge theoretical moduli problems,* Ann. Inst. Fourier (Grenoble) **56** no. 6, (2006) 1805–1826.
16. Bruasse L., Teleman A., *Harder-Narasimhan filtrations and optimal destabilizing vectors in complex geometry,* Ann. Inst. Fourier (Grenoble) **55** no. 3, (2005) 1017–1053.
17. Corlette K., *Flat G-bundles with canonical metrics,* J. Diff. Geom. **28**, (1988) 361–382.

© The Author(s), under exclusive license to Springer Nature Switzerland AG 2021
A. Zamora Saiz, R. A. Zúñiga-Rojas, *Geometric Invariant Theory, Holomorphic Vector Bundles and the Harder-Narasimhan Filtration*, SpringerBriefs in Mathematics, https://doi.org/10.1007/978-3-030-67829-6

18. Donaldson S.K., *A new proof of a theorem of Narasimhan and Seshadri,* J. Diff. Geom. **18**, (1982) 269–278.

19. Donaldson, S. K., *Anti self-dual Yang-Mills connections over complex algebraic surfaces and stable vector bundle,* Proc. London Math. Soc. (3) **50** (1), (1985) 1–26.

20. Dolbeault P., *Sur la cohomologie des variétés analytiques complexes,* C. R. Acad. Sci. Paris **236**, (1953) 175–177.

21. García-Prada O., Gothen P., Mundet i Riera I., *The Hitchin–Kobayashi correspondence, Higgs pairs and surface group representations,* arXiv:0909.4487.

22. García-Prada O., Heinloth J., Schmitt A.H.W., *On the motives of moduli of chains and Higgs bundles,* J. Eur. Math. Soc. (JEMS) **16** no. 12, (2014) 2617–2668.

23. Georgoulas V., Robbin J.W., Salamon D., *The moment-weight inequality and the Hilbert-Mumford criterion,* Preprint, ETH-Zürich, arXiv:1311.0410 (2013, last version (2019)).

24. Gieseker D., *On the moduli of vector bundles on an algebraic surface,* Ann. Math. **106**, (1977) 45–60.

25. Gieseker D., *Geometric invariant theory and the moduli of bundles,* Lecture Publication Series, IAS/Park City Mathematics Series v.00, (1994).

26. Gómez T., Sols I., *Stable tensors and moduli space of orthogonal sheaves,* arXiv:0103150.2001 (2001).

27. Gómez T., Sols I., *Moduli space of principal sheaves over projective varieties,* Ann. of Math. (2) **161** no. 2, (2005) 1037–1092.

28. Gómez T., Sols I., Zamora A., *A GIT characterization of the Harder-Narasimhan filtration,* Rev. Mat. Complut. **28** Issue 1, (2015) 169–190.

29. Gómez T., Sols I., Zamora A., *The Harder-Narasimhan filtration as the image of the Kempf filtration,* Proceedings of the Congress in honor to Juan Bautista Sancho Guimerá, Salamanca (2014).

30. Gothen P.B., *The Betti Numbers of the Moduli Space of Stable Rank 3 Higgs Bundles on a Riemann Surface,* Int. J. of Math. **5** no.6, (1994) 861–875.

31. Gothen P.B., Zúñiga-Rojas R.A., *Stratifications on the moduli space of Higgs bundles,* Portugaliae Mathematica EMS **74**, (2017) 127–148.

32. Griffiths, P., Harris, J., *Principles of Algebraic Geometry,* Wiley Classics Library, John Wiley & Sons, New York, (1978).

33. Grothendieck A., *Sur classification des fibrés holomorphes sur la sphére de Riemann,* Amer. J. Math. **79**, (1957) 121–138.

34. Grothendieck A., *Techniques de construction et théorèmes d'existence en géométrie algébrique IV: Les schémas de Hilbert,* Séminarie Bourbaki **221**, (1960–1961).

35. Halpern-Leistner D., *On the structure of instability in moduli theory,* arXiv:1411.0627v4 (2014, last version 2018).

36. Halpern-Leistner D., *Theta-stratifications, Theta-reductive stacks, and applications,* Algebraic Geometry: Salt Lake City 2015, (2018) 97–349.

37. Harder G., Narasimhan M.S., *On the cohomology groups of moduli spaces of vector bundles on curves,* Math. Ann. **212**, (1975) 215–248.

38. Hartshorne R., *Algebraic Geometry,* Grad. Texts in Math. **52**, Springer-Verlag New York, (1977).

39. Hall B.C., *Lie Groups, Lie Algebras, and Representations. An Elementary Introduction,* Grad. Texts in Math. **222**, Springer International Publishing Switzerland, (2004).

40. Hausel T., *Geometry of the moduli space of Higgs bundles,* Ph.D. Thesis, Univ. of Cambridge, (1998).

41. Hausel T., Rodriguez-Villegas F., *Mixed Hodge polynomials of character varieties, with an appendix by Nicholas M. Katz,* Invent. Math. **174** no. 3, (2008) 555–624.

42. Hausel T., Thaddeus M., *Generators for the cohomology ring of the moduli space of rank 2 Higgs bundles,* Proc. London Math. Soc. **88**, (2004) 632–658.

43. Hesselink, W.H., *Uniform instability in reductive groups,* J. Reine Angew. Math. **304**, (1978) 74–96.

44. Hilbert D., *Über die vollen Invariantensysteme,* Math. Ann. **42**, (1983) 313–373.

45. Hitchin N.J., *The self-duality equations on a Riemann surface,* Proc. London Math. Soc. **55** no. 3, (1987) 59–126.
46. Hitchin N.J., *Gauge theory on Riemann surfaces,* (M. Carvalho, X. Gomez-Mont and A. Verjovsky, editors), Lectures on Riemann surfaces: proceedings of the college on Riemann surfaces, Italy, (1989) 99–118.
47. Hoskins V., *Stratifications associated to reductive group actions on affine spaces,* Quart. J. Math. **65** Issue 3, (2014) 1011–1047.
48. Hoskins V., *Stratifications for moduli of sheaves and moduli of quiver representations,* Algebraic Geometry **5** (6), (2018) 650–685.
49. Hoskins V., Kirwan F., *Quotients of unstable subvarieties and moduli spaces of sheaves of fixed Harder–Narasimhan type,* Proc. London Math. Soc. **105** no. 4, (2012) 852–890.
50. Humphreys J., *Introduction to Lie Algebras and Representation Theory,* Grad. Texts in Math. **9**, Springer-Verlag New York, (1972).
51. Huybrechts D., Lehn M., *Stable pairs on curves and surfaces,* J. Alg. Geom. **4** no. 1, (1995) 67–104.
52. Huybrechts D., Lehn M., *Framed modules and their moduli,* Int. J. of Math. **6** no. 2, (1995) 297–324.
53. Huybrechts D., Lehn M., *The Geometry of Moduli Spaces of Sheaves,* Aspects of Mathematics E31, Vieweg, Braunschweig/Wiesbaden (1997).
54. Jansou S., *Le schéma Quot invariant,* J. Algebra **306** (2), (2006) 461–493.
55. Kapovich M., Millson J.J., *The symplectic geometry of polygons in Euclidean space,* J. Diff. Geom. **44** no. 3, (1996) 479–513.
56. Kempf G., *Instability in invariant theory,* Ann. of Math. (2) **108** no.1, (1978) 299–316.
57. Kempf G., Ness L., *The length of vectors in representation spaces,* In: Lønsted K. (eds) Algebraic Geometry. Lecture Notes in Mathematics, vol **732**, Springer-Verlag Berlin Heidelberg, (1978) 233–244.
58. King A.D., *Moduli of representations of finite-dimensional algebras,* Quart. J. Math. Oxford Ser. (2) **45** (180), (1994) 515–530.
59. Kirwan F., *Cohomology of quotients in symplectic and algebraic geometry,* Mathematical notes **34** Princeton University Press, Princeton, (1984).
60. Le Potier J., *L'espace de modules de Simpson,* Séminarie de géométrie algébrique, Jussieu, fév, (1992).
61. Maruyama M., *Moduli of stable sheaves, I and II.,* J. Math. Kyoto Univ. **17**, (1977) 91–126 and **18**, (1978) 557–614.
62. Maruyama M. (with collaboration of Abe T. and Inaba M.), *Moduli spaces of stable sheaves on schemes. Restriction theorems, boundedness and the GIT construction,* MSJ Memoirs **33**, (2016).
63. Markman E., *Generators of the cohomology ring of moduli spaces of sheaves on symplectic surfaces,* J. Reine Angew. Math. **544**, (2002) 61–82.
64. Meyer K., *Symmetries and integrals in mechanics,* Dynamical Systems (M. Peixoto, ed.), Academic Press New York, (1973) 259–273.
65. Mukai, S., *An Introduction to Invariants and Moduli,* translated by W.M. Oxbury, Cambridge Studies in Advanced Mathematics **81**, Cambridge University Press, Cambridge (2003).
66. Mumford D., *Geometric Invariant Theory,* Ergebnisse der Mathematik und ihrer Grenzgebiete, Neue Folge, Band 34, Springer-Verlag Berlin Heidelberg New York (1965).
67. Mumford D., Fogarty J., Kirwan F., *Geometric Invariant Theory,* Ergebnisse der Mathematik und ihrer Grenzgebiete (2) **34**, Springer-Verlag Berlin Heidelberg (1994).
68. Marsden J., Weinstein A., *Reduction of symplectic manifolds with symmetry,* Rep. Math. Phys. **5**, (1974) 121–130.
69. Nagata M., *Invariants of a group in an affine ring,* J. Math. Kyoto Univ. **3**, (1964) 369–377.
70. Narasimhan M.S., Seshadri C.S., *Stable and Unitary Vector Bundles on a Compact Riemann Surface,* Ann. of Math. (2) **82**, (1965) 540–567.
71. Newstead P.E., *Introduction to Moduli Problems and Orbit Spaces,* TATA Institute of Fundamental Research Lectures on Mathematics and Physics **51**, Bombay, Narosa Publishing House, New Delhi, (1978).

72. Nitsure N., *Moduli space of semistable pairs on a curve,* Proc. London Math. Soc. **62**, (1991) 275–300.
73. Ramanathan A., *Moduli for principal bundles over algebraic curves: I and II,* Proc. Indian Acad. Sci. (Math. Sci.) **106**, (1996) 301–328, and 421–449.
74. Reineke M., *Moduli of representations of quivers,* arXiv:0802.2147v1 (2008).
75. Riemann B., *Theorie der Abelschen Funktionen,* J. Reine Angew. Math (Crelle's journal) **54**, (1857) 115–155.
76. Rudakov A., *Stability for an abelian category,* J. Algebra **197**, (1997) 231–245.
77. Schiffmann O., *Indecomposable vector bundles and stable Higgs bundles over smooth projective curves,* Ann. Math. **183**, (2016) 297–362.
78. Schmitt A.H.W., *A universal construction for moduli spaces of decorated vector bundles over curves,* Transform. Groups **161** no. 2, (2004) 167–209.
79. Schmitt A.H.W., *Geometric Invariant Theory and Decorated Principal Bundles,* EMS Publishing House (2008).
80. Seshadri C.S., *Space of unitary vector bundles on a compact Riemann surface,* Ann. of Math. (2) **85**, (1967) 303–336.
81. Seshadri, C.S., *Fibrés vectoriels sur les courbes algébriques,* Astérisque **96**, (1982).
82. Shafarevich I.R., *Basic Algebraic Geometry,* Springer-Verlag Berlin, (1977).
83. Shatz S.S., *The decomposition and specialization of algebraic families of vector bundles,* Compositio Mathematica **35** no. 2, (1977) 163–187.
84. Simpson C.T., *Constructing variations of Hodge structure using Yang-Mills theory and applications to uniformization,* J. Amer. Math. Soc. **1**, (1988) 867–918.
85. Simpson C.T., *Higgs bundles and local systems,* Publ. Math. de l'IHÉS **75**, (1992) 5–95.
86. Simpson C.T., *Moduli of representations of the fundamental group of a smooth projective variety I and II,* Publ. Math. de l'IHÉS **79**, (1994) 47–129 and **80**, (1994) 5–79.
87. Terpereau R., Zamora A., *Stability conditions and related filtrations for (G, h)-constellations,* Int. J. of Math., **28**, Issue 14, 1750098 (2017) [34 pages].
88. Tits, J., *Reductive groups over local fields. Automorphic forms, representations and L-functions,* Proc. Sympos. Pure Math., Oregon State Univ., Corvallis, Ore., (1977), Part 1, 29–69, Proc. Sympos. Pure Math., XXXIII, Amer. Math. Soc., Providence, R.I., (1979).
89. Thomas R.P., *Notes on GIT and symplectic reduction for bundles and varieties,* Surveys in differential geometry **10**: A Tribute to Professor S.-S. Chern., (2006) 221–273.
90. Uhlenbeck, K., Yau, S.T., *On the existence of Hermitian–Yang–Mills connections in stable vector bundles,* Comm. Pure Appl. Math. **39**, (1986) 5257–5293.
91. Warner F.Q., *Foundations of Differentiable Manifolds and Lie Groups,* Grad. Texts in Math. **94**, Springer-Verlag New York, (2013).
92. Wells R.O., *Differential Analysis on Complex Manifolds, with an Appendix by Oscar García-Prada,* Grad. Texts in Math. **65**, Springer-Verlag New York, (1972).
93. Woodward C., *Moment maps and geometric invariant theory,* Actions hamiltoniennes: invariants et classification, Les cours du C.I.R.M. **1** num.1, (2010) 55–98.
94. Yang C.N., Mills R.L., *Conservation of isotopic spin and isotopic gauge invariance,* Phys. Rev. **96** no. 1, (1954) 191–195.
95. Zamora A., *GIT characterizations of Harder-Narasimhan filtrations,* Ph.D. Thesis, Universidad Complutense de Madrid, (2013).
96. Zamora A., *On the Harder-Narasimhan filtration of finite dimensional representations of quivers,* Geom. Dedicata **170** Issue 1, (2014) 185–194.
97. Zamora A., *Harder-Narasimhan filtration for rank 2 tensors and stable coverings,* Proc. Indian Acad. Sci. (Math. Sci.) **126** Issue 3, (2016) 305–327.
98. Zúñiga-Rojas R.A., *Homotopy groups of the moduli space of Higgs bundles,* Ph.D. Thesis, Universidade do Porto, (2015).
99. Zúñiga-Rojas R.A., *Stabilization of the homotopy groups of the moduli spaces of k-Higgs bundles,* Revista Colombiana de Matemáticas **52** no. 1, (2018) 9–31.
100. Zúñiga-Rojas R.A., *Variations of Hodge structures of rank three k-Higgs bundles,* Geom. Dedicata (to appear). arXiv:1803.01936.

Index

© The Author(s), under exclusive license to Springer Nature Switzerland AG 2021
A. Zamora Saiz, R. A. Zúñiga-Rojas, *Geometric Invariant Theory, Holomorphic Vector Bundles and the Harder-Narasimhan Filtration*, SpringerBriefs in Mathematics, https://doi.org/10.1007/978-3-030-67829-6

Printed in the United States
by Baker & Taylor Publisher Services